Evolution

by Frank B. Jevons

PREFACE

The object of this volume is to raise the question: if we accept the Theory of Evolution as true in science, how should it modify the thought and action of a man who wishes to do his best in this world? The question is necessary because we find that different and inconsistent conclusions on the point have been reached by men speaking in the name of science and speaking with authority. These differences are due not to anything in science, but to certain extra-scientific assumptions. To test the worth of such assumptions is the work of philosophy; and this volume is accordingly an essay in philosophy. Science is but organised common sense. Science and Religion both claim to deal with realities. The realism of common sense, therefore, the form of philosophy to which both seem to point, is that which is set forth here.

CONTENTS

PAGE

I. OPTIMISM

II. ILLUSION

III. PESSIMISM

IV. IDEALISM

V. THE REAL

VI. EVOLUTION AS THE REDISTRIBUTION OF MATTER AND MOTION

VII. NECESSITY

VIII. INSUFFICIENT EVIDENCE

IX. CONSEQUENCES

X. THE CHESS-BOARD

XI. THE COMMON FAITH OF MANKIND

XII. PROGRESS

XIII. EVOLUTION AS PURPOSE

XIV. CONCLUSION

APPENDIX. ON BISHOP BERKELEY'S IDEALISM

INDEX

EVOLUTION

I.

OPTIMISM

Innumerable writers at the end of the nineteenth century have reviewed the changes which in the last fifty years have come over the civilised world. The record indeed is admitted on all hands to be marvellous. Steam, electricity, machinery, and all the practical inventions of applied science have added enormously to the material wealth, comfort, and luxury of mankind. Intellectually, the bounds of pure science have been vastly enlarged; and the blessings of education have been extended to the poorest members of the community. Philanthropic and religious activity manifests itself in a thousand different organisations. We are never tired of repeating, that changes which in the first half of the century would have been pronounced impossible and incredible, at the end of the century are accomplished facts.

But amongst all these changes one is almost universally overlooked, and that the most characteristic, the most remarkable, and the most important: the face of civilisation has come to be illumined by hope. Great as is the progress of the last fifty years, we count it as nothing compared with that which is in store for us. To the discoveries of science it is felt that no bounds can be set; what a day may bring forth in the way of the extension of man's control over the forces of Nature, what secrets of Nature the chemist in his laboratory may light upon at any moment, no man can surmise, but everyone is confident that things will be discovered as marvellous to us now as the telegraph and telephone to our predecessors of the pre-scientific age. In the treatment of political and social questions the same deep-seated conviction prevails that progress can and will be made: the conditions and causes of poverty can be ascertained by patient study, and when ascertained can be dealt with. The laws of physical health and cleanliness have not refused to reveal themselves, nor are moral health and cleanliness without their laws. In fine, if the best energy of the age is everywhere devoted to the increase of knowledge, the advancement of morality, and the diffusion of comfort, it is because everywhere there is hope. In the social as in the individual organism hope raises the tide of life, increases vitality, and stimulates the system. Hence this general discharge throughout the nervous system of society, manifesting itself in the vigour and energy with which all schemes for improvement are taken up and carried out. That discoveries will be made and progress effected is as certain as that gold is to be found in a goldfield; the only practical question is, By whom? Who is to be the lucky man?

To us who have witnessed the advance which has given rise to this universal hope, the hope itself seems so reasonable and so justifiable that we are apt to overlook the fact that it is without parallel in the history of mankind. Never, of course, has any generation of men imagined its own lot perfect; all have had their ideals, and all have believed their ideals to be true. But whereas we place the realisation of our ideals in the future, all previous generations have placed it in the past: the Golden Age till now has always been regarded as the starting-point of man's history, not its goal. All races have looked back with pride upon a heroic past; all mythologies tell of the better and brighter lot that was in the beginning man's; all poets sing of the brave days of old; all fairy tales begin with "once upon a time." The historians of Greece and Rome discovered no progress in the history of their countries, but only degeneration from the patriotism and simplicity of earlier times, or at best a series of changes making its round like the circle of the year's seasons. The philosophers of Greece are mainly occupied, when they deal with sociological questions, with the causes of corruption and decay of constitutions; and, if they frame ideal constitutions, they intend them to be final; they do not imagine them to have any possibility of growth. In modern times the same tendency has been equally manifest. Political revolutions have always aimed, not at introducing a new, but at restoring an old state of things: the actors in the French Revolution even dressed and posed as ancient Greeks and Romans. In philosophy, civilisation, as being artificial, has been regarded as a degeneration from a "natural" state of man which was at once primitive and perfect.

In the individual, optimism may be dismissed as a mere mood, or as a tendency to cheerfulness not based on any rational estimate either of the future or of the past. But when a whole generation of men, when, indeed, the whole civilised world, looks to the future, not with careless levity, but with the calm assurance of confidence in the progress that is and is to be, we cannot dismiss its optimism offhand. Astonishing as it is, that the world as it grows older should grow more hopeful, there are good reasons for the fact.

The child's estimates of distance, magnitude, and importance differ from those of the adult. The estimates, however, persist in memory, and we have all discovered, on revisiting familiar scenes of childhood, how exaggerated our childish estimates were when compared with the actual facts. It is this exaggeration of memory, this illusion of the mind's eye, that psychologically is the foundation of the tendency to idealise the past. To us as children the exploits of our elders were marvellous in our eyes; and they remain as marvels in the memory, as marvels, however, which, as all marvels do, belong to the past. The past becomes the wonderland in which were performed the great deeds, not only of our fathers' time, but of the old times before them. The past becomes the poet's treasury, from which he produces things new and old--the abiding-place of all things good and great and beautiful which are not, but ought to be, and therefore once were.

To measure progress, as indeed to measure any movement and determine its rate and direction, some fixed points are necessary. As long, therefore, as there is no contemporaneous record of events, fixed in writing, there is no possibility of checking the laudator temporis acti and of reducing the unconscious exaggerations of his memory to their due proportions. But even if there were, in the lowest stages of culture the rate of progress is too slow to be perceptible at the time. In the beginning man is at the mercy of his environment: it is only when he has learnt to modify it to his needs that progress begins to move. And by the time that man has passed from savagery to barbarism, and has emerged from barbarism to civilisation, the conviction that the present and the actual are things of naught as compared with the ideal past, is too intimately inwrought with his religion, his mythology, his philosophy, and the accepted history of his race and its heroic origin to allow him to see facts as they are, or to divine the true trend of human affairs. Further, there is a very practical reason for his looking with suspicion and not with confidence on social changes. It is only as the result of a long course of slow evolution that society has attained to a condition of fairly stable equilibrium. In the beginning society may be compared to a man hanging on for bare life, with a precarious foothold, to the face of a sheer cliff: when the least movement may prove fatal, all movement is dreaded. Thus the characteristic of all early societies is that they are impeded by "the cake of custom" and rigid with the immobility of conservatism.

To those who hold that experience mechanically impresses itself upon the mind and so automatically expresses itself as truth, it must appear somewhat strange that mankind should have advanced for thousands of years without knowing that they had progressed; and still more strange that it was not as an induction from experience, but on a priori grounds that they arrived at the conclusion. Yet so it was. The mere contemplation of the rise and fall of empires no more suggested the presence and persistence of a constant tendency to progress than the mountainous wave which threatens to engulf the ship suggests that the sea-level is a scientific truth. But when Darwin established his theory that man was descended from the brute, all was clear: it became certain a priori that the long history and "pre-history" of man must have been one of progress and advance. When the descent of man was established, his ascent came to be studied, and human evolution was seen to be synonymous with progress. Savages were seen to be the nearest existing representatives of primitive man, and there was an end to the idea that the primitive state was perfection. The comparative method, once applied to the study of mankind, was able to set side by side examples of savagery, barbarism, and civilisation, which illustrated every step in the process of the evolution of society, and showed that, though the forms of society may fluctuate as do the waves of the sea, society itself is steady in its advance and progressive in its evolution. This conclusion, which at first was a deduction drawn from the animal descent of man, has now the independent support of an enormous amount of evidence. The existence of a Stone Age, paleolithic and neolithic, of a Bronze Age and an Iron Age, and the succession of those ages in the order named, are established facts of science. That the culture of nomad peoples is lower than that of pastoral tribes; that pastoral tribes advance in culture when they become agricultural; that agriculture, implying settled habits and fixed homes, leads to the foundation of cities and the formation of civic life; that the city-states of the ancient world give way to the nation-states of modern times: are all accepted facts, bridging the apparent chasm between civilisation and savagery, and demonstrating the action of the law of continuity in the evolution of society.

But, it will be observed, all these facts and arguments taken together only prove what has been--not what will be. They show that from a level little higher than the brute man has attained to what he is; but is this enough to guarantee his continuous rise? In other words, have we reached the real source of that universal hope which, as we have said, is characteristic of this stage of man's evolution? The bark of man's destiny hitherto has been wafted by a favouring and a steady gale, and it is natural enough for the unreflecting to take it for granted that the wind will always set from the same happy quarter. But the question will obtrude itself whether we are justified in the presumption.

If man shaped his own course, we might at least say that there was no reason why he should not continue to steer in the same direction as hitherto. But the most remarkable lesson that sociology has to teach us is that the course which he has followed so continuously has not been of his own steering. As we have already seen, man until this present generation has uniformly kept his eyes fixed on the quarter from which, not to which, he has imagined himself to be travelling, and, like a reluctant emigrant, has lamented the increasing distance between him and the happy shore from which he sailed. Or, to change the metaphor, society is an organism. Like all organisms, it starts as a relatively structureless mass; then, in accordance with the principle of the division of labour, different functions come to be performed by different parts; thus special organs are developed for the performance of special functions; division of labour further implies co-operation of the various organs and the development of the necessary means of communication and connection. All this is necessary for that evolution of society which we call progress; and of all these changes in the structure of society but few were ever intentionally planned by man. Mr. Herbert Spencer has familiarised this generation with the idea that the foreseen consequences of any intended change are insignificant as compared with the consequences unforeseen and unintended. Hence the general rule that the structural developments on which the evolution of society depends are but rarely the result of the coercive and conscious changes effected by government: in practically all cases they are the unintended consequences of the spontaneous actions of individuals aiming at something else and unconsciously promoting the evolution of society. So, too, the animal organism is made up of living units, each of which unconsciously performs the part necessary to be played by it, if the organism is to live; and each unit, unconsciously again, even modifies the part it plays, in order to promote the changes which constitute the evolution and the progress of the organism.

We must therefore dismiss the idea that the progress of mankind and the evolution of society have been planned by man or are due to his design; and we must recognise the presence in human affairs of some unseen, impelling power which is continually guiding them to good issues and shaping them to ends not even rough-hewn by men. This power, it is evident, must be one not limited in its action to the social organism, but manifesting itself in animal organisms also, since there also it produces similar results. That power, we need hardly say, is to be sought in "the struggle for existence": wherever organisms are in excess of the means for supporting them, competition for food, for life, must ensue; and in this case the battle is to the strong, the race to the fleet. But of course strength is a relative term: what in some circumstances is a source of strength, in others may be a cause of weakness; and, generally, the very qualities which in some cases are of the highest value may in others be useless to their possessor. It is therefore the creature which possesses the particular kind of superiority required by the circumstances in which it finds itself, which is the creature that is likely to fare best, and is most likely to survive in the struggle for existence. But, further, the circumstances tend to produce the very superiority which they require: they ruthlessly reject and condemn to destruction every organism which fails to satisfy their requirements, thus leaving the field in possession of those organisms which have the required superiority. The next generation, therefore, is bred not from chance parents, but from parents which have been selected, by natural causes and the force of circumstances, as carefully as by the breeder who wishes to produce a prize animal. Every successive generation thus must be superior to that which preceded it. Advance is the very breath of every organism's being, the condition without which existence is impossible. To the talents which it has, every being must add other talents, or be cast out into the darkness of non-existence; whereas to the good and faithful servant who exercises all the powers entrusted to him even wider rule is given. Neither this world nor the next is for the idle or for the stupid. The intelligence must be alert to detect the slightest element of possible superiority, and the will resolute to work it to the utmost of its worth. Man must be wise in his generation; and the wise man makes friends even with the mammon of unrighteousness, and that quickly.

If, then, it is by the perpetual and strenuous exercise of all its powers that an organism achieves the degree of superiority which is its contribution to the universal work of progress, it follows that "the performance of every function is, in a sense, a moral obligation," and that "the moral man is one whose functions are all discharged in degrees duly adjusted to the conditions of existence." Here, as elsewhere, the individual, to exist, must comply with the conditions of existence; and progress consists in more perfect compliance with the conditions. There is, however, a difference between the highly evolved organism, man, and the less complex organisms; between animal and human evolution; between biological and moral progress. In the case of the lower and simpler organisms, the creature is prompted simply and safely by its emotions to the performance of those functions on which its existence and the evolution of its species depend. But the evolution of man has been so rapid in its later stages, the social environment which he has himself created is so different from the circumstances in which he originally found himself, that his adjustment to his environment has become, so to speak, much looser, and consequently it is now no longer the case that actions in themselves pleasant are also necessarily beneficial in their consequences to the individual and to society. Moral progress, therefore, will manifest itself in the readjustment of man to his altered conditions. The consequence of that adjustment, when complete, will be that actions which are right--that is, are beneficial to the individual and to society--will always be pleasurable, not only in their consequences, but also immediately and in themselves. To this ideal, when all men will delight always in the thing that is right, and when all have attained to a height of morality now reached only by the few, man is being slowly but surely urged by the force which is the motive power of all evolution, the struggle for existence, regulated by the law which directs all progress, that of the survival of the fittest.

Here, then, we have the reason of the hope that is characteristic of our generation; here the foundation of the calm confidence with which we count on the continuance of progress as a thing assured us. It is not merely that progress has been made in the past, that the gale hitherto has steadily blown us on a favourable course. We have learnt that it must of necessity always blow from the same quarter. Man's course is not dependent on man's fitful will: the wind and waves obey not him, but the Power which directs all evolution, and "our strength in ages past" is shown by science to be "our hope in years to come."

II.

ILLUSION

It seems, then, according to the optimistic view set forth in the previous chapter, that Evolution is necessarily Progress, and progress is movement in the line of our moral aspirations produced ad infinitum. The changes that are and always have been taking place are and always have been changes for the better; the forms of existence which incessantly succeed one another necessarily develop from lower to higher, from good to better. And this conclusion is not a matter of religious faith, but of scientific necessity. The only forces and causes that it presupposes are those which we see and feel at work every day around us. For the reconstruction of the past history of the earth's surface, geology only requires to assume the operation during infinite past time of those agencies which at this moment may be seen to be slowly changing the face of the earth. The cooling of the earth's surface follows the same laws, and can be calculated with the same certainty as the cooling of a red-hot poker. The law of gravitation, which determines the movements of the heavenly bodies, is equally exemplified in the fall of an apple to the ground. In fine, the universe consists of bodies of matter in motion; the movements which occur within the range of human observation are sufficient to enable us from them to calculate the paths which they follow when they pass beyond our ken, and the correctness of our calculations is demonstrated when they reappear at the time and place predicted. The chemist recovers on one side of his equation every atom which the other side requires him to account for. The stars in their courses confirm the calculations of the astronomer. Matter is in perpetual course of redistribution, and the same everlasting laws which determine the forms into which it is incessantly being redistributed necessarily determine that those forms shall perpetually improve.

This optimistic view of evolution has met with general welcome, but on very different grounds in different cases. Believers in Divine Providence have eagerly greeted it as a startling and irresistible demonstration that their belief in a Providence over-ruling all things for good was true. No suspicion here was possible that the argument had been sophisticated by those with whom the wish was father to the thought. By science the testimony of science could hardly be impeached; and here was science on independent reasonings of its own, starting from purely materialistic ground, compelled by the force of its own arguments to bear witness to the truth which religion had so long proclaimed on the strength of faith alone. To this generation a sign had indeed been given.

On the other hand, the optimistic interpretation of evolution was welcomed with equal ardour by those for whom it removed the last difficulty they had in believing that there was no God. Hitherto the deeply rooted desire to believe that, in spite of all appearances to the contrary, good must triumph ultimately, and right-doing never be confounded, had seemed to necessitate belief in a righteous God. But now the necessity for any such assumption was done away with: the perpetual triumph of the good was a necessary aspect or expression of the mechanical action of particles of matter upon one another, as much as the law of gravitation itself, and based on exactly the same kind of evidence. From this it followed that religious belief was but a passing phase in the process of evolution, useful enough as long as the real evidence for our faith in the good was unknown, but destined to dwindle to a mere rudiment and survival as fast as men become capable of seeing the truth of the matter, and of realising that religion is superfluous because it can offer nothing that is not independently assured by science. At the same time and in the same way the hope of future blessedness is brought down from the unsubstantial clouds of an imaginary heaven to the solid ground of a materialistic science, which never travels beyond the evidence of the senses.

Since, then, minds, which differ otherwise so much, are agreed that the optimistic interpretation of evolution is the true one, it seems not unreasonable to ask each how far they are prepared to push their optimism. We will ask the one side whether the reason why they believe in the goodness of God really is that, as a matter of fact, they see that good is incessantly triumphant around them, and triumphant as a matter of absolute necessity. Surely whether we consider what we daily see of life, or whether we consider the struggle with evil in our own souls, it is a mockery to say that good invariably triumphs here and now; and there must be illusion in the argument that would prove that it does. Could an argument that is based on the assumption that matter and motion are the only realities issue in anything but illusion when extended to spiritual experience?

To the other side we may put the question somewhat differently. It is agreed that all the many changes which are incessantly taking place in the universe, and which, added together, constitute what is called the cosmic process, are incessantly and inevitably working for good, and themselves are always rising from good to better. But what of the Force, or Power, or Cause, or Reality which underlies them and of which they are the manifestation? May we infer that because they are good, it is good? That if the fruits are good, the tree must be good also? To this the reply will be that it is the manifestations which we know; they alone are known to us; they alone can be known to us. That which underlies them is not manifest; and that which is not manifested to us obviously cannot be known to us: it is the Unknowable. Obviously, therefore, it is impossible for us to say whether it is good or not. To affirm and to deny that it is good would both equally be to profess knowledge of the unknowable. Religion may profess--and, indeed, all religions have professed--to possess this inconceivable and impossible knowledge. But religion is not science.

On this view, then, there are limits to the optimism of evolution: to apply the term "good" to that which manifests itself as the cosmic process in evolution is mere illusion. But this raises a further question: If it is unmeaning to call the Unknowable Reality good, what precisely is the meaning and value of the term "good" when applied to those forms in which the Unknowable manifests itself to us?

To begin with, it is clear that if everything has been evolved, then our moral aspirations also are the products of evolution. It is they, indeed, that distinguish man from the brute; but even of them the law of continuity holds good: we can see not only how in man the virtues have been developed by civilisation, but we can trace the germs of conscience in that civilised animal the dog, as we can certainly see maternal affection, devotion, and self-sacrifice in the fiercest of undomesticated animals. In other words, the struggle for existence is waged better in co-operation than by individual effort; co-operation implies the subordination of individual impulse to the interests of the species or society; and such subordination, taking different forms in different stages of social development, is what we call virtue.

In the next place, the theory of evolution is built upon the ancient truth that nothing abideth long in one stay. Matter and motion are in perpetual course of redistribution, entering into countless combinations, and assuming innumerable forms, which succeed each other like the waves of the sea, and like them are no sooner formed than they are gone. It follows, then, on this showing, that our moral aspirations are as transitory as other products of evolution. Indeed, as we look back over the pages of history we can see them always changing before our eyes--what is approved by savages is disapproved later; the virtues of the military stage of social development give way to those fostered, by the industrial organisation of society. In a word, our moral aspirations, being the outcome of evolution, have neither the permanence of matter and motion which are everlasting and indestructible, nor the reality which is the attribute of the Unknowable alone.

If any confirmation of this conclusion were required, it would be found in the fact that only a living, conscious being can entertain moral aspirations, or desire the good, or hunger and thirst after righteousness. And life and consciousness are but transitory phases of evolution. The earth's crust, the geologic record, testifies to the former existence of fauna now extinct. The science of heat makes it certain that the earth must cease to be habitable for any form of life; and with the extinction of consciousness, good and the desire for good, right and the striving after right, will be no more: matter and motion, brute matter and blind forces, knowing nothing of good or evil, will resume their ancient, desolate domain.

If, pursuing the same train of thought, we ask what meaning the optimistic evolutionist puts upon the word "good," we shall see that, according to him, the distinction between good and bad is one that applies, and can only apply, to certain moments in the process of evolution, but not to the process as a whole, just as we have already seen that according to the optimistic evolutionist the distinction does not apply to the Unknowable Reality of which the process of evolution is a manifestation. The law of life is laid down to be the struggle for existence, with the consequent survival of the fittest. In the struggle, that is good which is struggled for, viz. existence; and that conduct, in man or brute, is good which conduces to success in the struggle and enables the organism to maintain its existence. This can only be done by the adaptation of the organism to its environment, of the constitution to the conditions. It follows, therefore, that "good" is a purely relative term: it is only applicable with reference to organisms, and even in their case only to success and whatever contributes to success in the struggle for existence. But to the cosmos before the struggle for life begins, and after life and its struggles have relapsed into the insentience of unconscious matter, the term cannot be applied. Matter and motion, which exist before and after life's appearance, are everlasting and indestructible. Their existence is assured, and implies no struggle. They are eternal, organic life compared with them is momentary. The portion, then, of the cosmic process which can be spoken of as good is infinitesimal compared with the whole. Save for the brief moment during which organic life exists, it is as illusory to speak of the cosmic process as good as it is to apply the term to the Unknowable.

But if so much of our optimistic interpretation of evolution has proved to be an illusion which consists in the simple fallacy of using the word "good" in connections in which it has no meaning, can we hope to rescue the very small fragment that remains? Perhaps we may argue, that since that is good which conduces to human existence, the whole of the cosmic process up to now, having paved the way and prepared the earth for man, must be good. Thus at one stroke we seem to regain half at least of the territory we have lost. But it is only seeming, once more illusion, for the cosmic process which has prepared the earth for man's existence has also prepared it for his destruction: his good, his existence, and his destruction are equally indifferent to it. This conclusion is confirmed by the reflection that to regard the cosmic process as giving any consideration to man would be to ascribe purpose, consciousness, a knowledge of good and evil, and a preference for good, to the Unknowable of which the cosmic process is the continuous manifestation.

It is therefore mere illusion to imagine that evolution necessarily tends to good: it is absolutely indifferent to it. And as we must judge of the parts by the whole, we must conclude that human evolution follows the same laws as evolution in general. The steps in human evolution, like those in evolution at large, are not progress, are not changes working to a good end, but merely changes. Evolution is not progress, but mere change, as far as good and evil are concerned, a mere marking of time, or at most a series of movements in which advance and retreat cancel each other in the long-run.

At the same time, the evolution theory enables us to see plainly a cause at work which would inevitably produce in human minds the illusion that existence is good. Just as any species of animals which found a pleasure in actions ultimately entailing the destruction of the species would be condemned to extinction, so too only those varieties of the genus homo could survive in whom the conviction of the goodness and desirability of existence was strong enough to call forth the activities on which existence was dependent.

The optimistic interpretation of evolution is based on the "struggle for life" theory that "existence" sums up the good for which man struggles; and we have sought to show that the optimism which is based on this assumption must result in the conclusion that progress is an illusion. Some readers, however, may hold that mere existence is not the only good that man is capable of struggling for.

III.

PESSIMISM

"The prospect of attaining untroubled happiness, or of a state which can, even remotely, deserve the title of perfection, appears to me to be as misleading an illusion as ever was dangled before the eyes of poor humanity. And there have been many of them."[1]

The theory which sees in evolution nothing but the redistribution of matter and motion leads to an optimistic view of things which on examination proves to be a misleading illusion. From illusion to pessimism is but a step.

The facts on which the theory of organic evolution is based are two. The first is that no two individuals of any species are born exactly alike; and that of two different individuals one must be superior to the other, i.e. better fitted to survive under the conditions then and there prevailing. The next is that parents transmit their qualities to offspring; and the superiority of superior parents is thus transmitted and accumulated from generation to generation. Organic evolution, therefore, consists in more and more perfect adaptation of the organism to the environment. And this adaptation is effected by the physical destruction of those creatures which are weakly and not adapted to cope with the environment.

According to the theory that evolution is progress, the progress or evolution of humanity obeys the same laws, is impelled by the same forces, and follows the same line as the evolution of organisms in general; and consists accordingly in increasing adaptation to the environment. Imperfect adaptation manifests itself whenever a man's impulses or desires move him to perform acts which are immediately or eventually prejudicial to his own or to society's existence. Adaptation will be perfect when all acts which are necessary for the existence of the individual and of the society are pleasant in themselves--when not only going to the dentist's will be a duty, but the extraction will be a pleasure desired for its own sake.

Though Mr. Huxley maintained that it was a misleading illusion to lead people to expect any such state of untroubled happiness, he was far from denying that progress has been made in the past by man, or from despairing of further progress in the future. But progress does not, according to him, consist in adaptation to environment; it is not effected by means of the struggle for existence; it neither obeys the same laws, nor is impelled by the same forces, nor follows the same lines as organic evolution in general. Nor does it consist in the substitution of personal pleasure for a sense of duty as the motive of action: on the contrary, it consists in a fuller and fuller recognition of the claims of others.

The idea that evolution means progress, and by its very nature necessarily results in perfection, owes much of its popularity to the fallacious interpretation given to the phrase "survival of the fittest." In any scientific use of the phrase, "fittest" simply means "fittest to survive." But in popular usage it is supposed to mean "ideally or ethically best." But the fittest to survive are not necessarily the ideally best: they are, scientifically speaking, simply those best adapted to the circumstances and conditions under which they live. And the circumstances and conditions, the environment, may or may not be favourable to the survival of the ethically or aesthetically best: they may be favourable to the growth of weeds and to the destruction of beautiful flowers, in which case the cosmic process will wipe out the beautiful flowers, and the movement of evolution will be aesthetically retrogressive, not progressive.

Adaptation to environment, therefore, is no indication or test of progress, or of what is good or right or true or beautiful. Everything that exists is shown, by the mere fact of its existence, to be adapted to its environment. If, therefore, such adaptation is evidence that the thing is ideally satisfactory, it will follow that whatever is, is right. At the same time, our conception of right and good will be emptied of all meaning: a "right" or "good" thing will simply mean a thing which exists. The epithets will simply predicate existence, not a quality; and consequently we shall have to call the successful villain and the prosperous traitor good, and their methods right. They have adapted themselves to their conditions, and have flourished in consequence.

Adaptation to environment could only mean progress provided that the environment was uniformly such as to favour the survival of those alone who were ideally fit to survive. But it is not: instances are not uncommon in which organisms, having attained to a certain degree of complexity and heterogeneity of structure, subsequently, as a consequence of adapting themselves to their environment, lose it and revert to an earlier stage of development, relatively simple, homogeneous, and structureless. Such reversion or regressive metamorphosis is as much a part of the organism's evolution as its previous progressive metamorphosis; and progress and regress both are equally the result of adaptation to environment. Further, though reversion and regress may now be only occasional, it is certain that as the earth cools down they must become universal: the altered conditions of temperature, etc., will allow only the lower forms of life to survive, and will eventually extinguish even them.

As regards organic evolution in general, then, the struggle for existence and the action of the environment do not necessarily tend to result in progress. As regards the evolution of man in particular, Mr. Huxley went further and maintained that they were absolutely inimical to human progress, which has been effected, not because, but in spite of them, and is the result not of obeying the cosmic process, but of defying it.

The qualities which brought success in the struggle for existence to man as an animal were rapacity, greed, selfishness, and an absolute and cruel indifference to the wants and sufferings of others. On the gratification, at all cost to others, of his animal desires, his animal existence depended: it was the "ape and tiger" within him that made him victor in the struggle for existence; it was the environment that imposed this as the condition of success.

The qualities which make man a human being are tenderness, pity, mercy, compassion, self-sacrifice, and love. It is in their growth--the "ethical process"--that human progress consists, and not in the ruthlessness by which the cosmic process effects the evolution of other organisms. These qualities--human and humane--do not make for success in the struggle for existence. They are not adapted to the environment provided by Nature. Their owners were not the fittest to survive, and consequently paid the penalty--physical destruction--as far as the cosmic process could exact it. If the struggle for existence and the action of the environment have not succeeded in keeping man down to the level of the brute, it is because man has deliberately set himself to oppose the cosmic process and the blind forces, knowing nothing of right and wrong, pity or love, by which it effects the evolution of the brute. The struggle for existence is fatal to the development of the qualities which are peculiarly characteristic of humanity, and man accordingly has suspended the struggle for existence. In place of warring with his fellow-man, he has begun to co-operate with him. He has learnt to some extent to postpone the gratification of his own wants to the satisfaction of those of others. He no longer destroys the weakly, the sick, the helpless, the useless, or even the criminal; and, if the environment threaten their destruction, he sets to work to alter the environment. Man no longer seeks to conquer Nature by obeying her: he studies her forces in order to command them to his will. Adaptation to environment is the implement by which she shapes human evolution to ends that are not his ends; he wrests the weapon from her hands, and by adaptation of the environment undoes her work, fosters the growth of those qualities which tend towards his ideal, and does away with the conditions which harbour ignorance and error, selfishness and sin.

Human progress, then, consists in perpetual approximation to the ideals of charity, love, and self-sacrifice. Life is exhibited as a struggle against evil, against the ape and tiger within us which we inherit from our ancestor--the brute. The evil is real, the struggle is hard but worthy, and not the less worthy because it is not directed to our personal happiness and gratification. "The practice of self-restraint and renunciation is not happiness, though it may be something much better."[2]

Thus far this criticism of life, though stern, is not pessimistic. On the contrary, in it man seems to have recovered the freedom of action and the power of independent judgment which, as the mere product of the cosmic process, he could not enjoy according to the optimistic theory. If life is a struggle, at any rate man can fight the good fight, if he will; and he can judge for himself which is the higher, the adaptation to environment which puts man on a level with the ape and tiger, or the adaptation of environment which, for the sake of his ideals, sets him in conflict with the cosmic process.

It is when we proceed to conjecture the issue of the struggle, as thus stated, that pessimism begins to invade us. However valiantly man may fight, whatever temporary victories he may gain here or there, his defeat in the end is inevitable: the same cosmic forces which, working through him, have won him his trifling victories have preordained his ultimate destruction. As far as it is possible for science to forecast the future, it is certain that in the end man will fall a victim to his environment, and join the other extinct fauna of the earth. With him the ethical process ceases; with him perish the hopes, the aspirations, and the ideals for which he strove as being of greater worth than aught that evolution, the redistribution of matter and motion, could offer or produce.

If this were all, the picture would be sufficiently gloomy: man alone in the universe, surrounded by forces which act without regard to good or evil, without sympathy or heed for right or wrong, indeed, with the effect of impartially extinguishing both in the end. But it is not all. As the conditions grow more and more unfavourable to man's existence upon earth, as the margin of the means of subsistence contracts, and the presence of universal want increases, the ape and tiger in man will begin to assert themselves once more. In the face of starvation, the instinct of self-preservation will become imperious. Once more, as in the earliest days, man will live by rapacity, cruelty, and selfishness alone. Before man yields possession of the earth to the brutes, he will himself revert to brutishness. The puny barriers behind which man has for a moment sheltered himself from the action of the cosmic process, and nursed the feeble flame of those aspirations after higher things which distinguish him from the brute, must inevitably be swept away by the restless and relentless tide of insentient matter, perpetually redistributed by aimless motion, which constitutes the cosmic process.

The pity of it is that the process of evolution should require not merely man's physical destruction, but his moral destruction also; that the ruin of his body should be preceded by the ruin of his soul; that in his regressive metamorphosis he should be compelled, by the struggle for existence and the instinct of self-preservation, to play the traitor to one after another of his ideals of tenderness, of pity, and of love. The fittest to survive will be those who are most completely adapted to the altered environment, who are resolved to succeed in a struggle for existence in which success can be obtained by brutishness alone. The least fitted to the new conditions, and the first to perish therefore, will be those with whom self does not come first. With their destruction the competition between their less scrupulous survivors will become fiercer and still more cruel. And this process will be repeated again and again, each generation transmitting cunning and cruelty intensified to the next. Our great cities already breed men degraded below the level of the lowest savages known to us, but even they can give us but little idea of what the struggle for existence will yet produce from the ruins of civilisation in the course of the Evolution of Inhumanity.

While proclaiming that "the ethical process is in opposition to the principle of the cosmic process, and tends to the suppression of the qualities best fitted for success in that struggle," and that at the best the ethical process can maintain itself only for a relatively short time, "until the evolution of our globe shall have entered so far upon its downward course that the cosmic process resumes its sway; and, once more, the State of Nature prevails over the surface of our planet," Mr. Huxley held that our duty lay "not in imitating the cosmic process, still less in running away from it, but in combating it."[3] "Cosmic nature is no school of virtue, but the headquarters of the enemy of ethical nature," and though we know that the enemy's triumph must be complete, that the defeat of the good cause is preordained, that we and ours must be annihilated, we must remain at our posts, fighting to the end without hope.

It seems, then, that man possesses two kinds of knowledge: he knows to some extent what is, and to some extent he knows what ought to be. And both kinds of knowledge are equally valid. He judges that a thing is, and he judges also that a thing ought to be. Both judgments are equally true, but apparently both are not equally final, for if man judges that what is, ought not to be, he is impelled to alter what is, so that in the end the thing that ought to be is also the thing that is. The judgment of what ought to be, the ideal, is thus proved, or rather made, to be the finally correct one. On the other hand, if man is defeated in his attempts to adjust the things that are to his judgment of what they ought to be, he does not acquiesce in his defeat; he refuses to accept the result as final; the end of the matter is not there; things are not what he strove to make them, but they ought to be. What is has nothing to do with what ought to be.

But what ought to be may make a good deal of difference to what is.

The ethical process, in its conflict with the cosmic process, may not in the end prove victorious; but that makes no difference to the fact that it ought to be victorious. It is this deep-seated conviction which made Mr. Huxley say that we must declare war to the last against cosmic nature, the headquarters of the enemy of ethical nature. The victory of the enemy may be certain, but it will none the less be wrong; it may be permanent, but as long as it lasts it will be wrong. If matter and motion are eternal and indestructible, morality is equally everlasting and immutable. Unless this is so, unless the triumph of the cosmic process is wrong, once and always, why are we called upon to endure sorrow and pain and suffering rather than submit to it? Our judgment that it is wrong is as independent of time as is our judgment that particles of matter gravitate towards one another. We have no reason for believing that the latter will continue to be true for a longer time than the former. Indeed, if matter and motion, having achieved their victory over the ethical process, were then and there to be annihilated, their victory would continue to be wrong, though they had ceased to be. Right may triumph or wrong may triumph, but right is right and wrong is wrong for evermore. It is vain to tell us in the same breath that we must stake our all upon our moral judgments and that our moral judgments are not to be relied on. Every impeachment of their validity is an invitation to us to give up the struggle against the enemy of ethical nature. And if we are really resolved to fight the good fight and quit ourselves like men, we thereby affirm that our moral judgments are at least as valid as our judgments on matters of fact, and that, if our knowledge of what is is true objectively, our knowledge of what ought to be has in it at least an equal element of objective truth.

If, then, the cosmic process is real and objective, in so far as it is a perpetual manifestation of the Unknown Reality which underlies all things, then the ethical process, having the same reality and objectivity, is also a manifestation of the Unknowable. The perpetual redistribution of matter and motion is not the only way in which the Unknowable manifests itself to men: it also gives a shape to itself in the form of the highest and purest aspirations of which man is conscious within himself. It might seem, therefore, at first sight as though a mere dispassionate consideration of the actual facts of life, quite apart from any religious presuppositions or presumptions, forced us at last into the presence of a God, the source and author of all goodness. But, in the first place, those who hold to the dogma of Agnosticism, that what underlies things as they are known to us is the Unknowable, cannot admit that we know or can find out whether the Unknowable is good or bad. Induction, the logical method to which science owes so many of its discoveries, and by which we proceed from the known to the unknown, does not avail us here. No logical method could discover what is not merely unknown, but absolutely unknowable.

In the next place it is reasonable enough that those who begin by believing in a Divine Providence should also believe that right will triumph in the end, if not in the world as it is manifested to us now and here, in space and time, then in that real world, that kingdom of heaven, of which this world is but an imperfect manifestation, or to which it is but a distant and slowly moving approximation. For those, however, who refuse to assume the reality of a Divine Providence the case is different. They base themselves on facts of experience: they observe that to some small extent what ought to be tends to substitute itself for what is, thanks to the action of man exclusively, and not to any inherent tendency to good in cosmic nature, but rather in spite of the resistance to good caused by the necessary action of the mechanical laws of nature. From their observation of the conditions under which man has succeeded in modifying what is into what ought to be, they forecast the extent to which that process may be carried in the future; and their conclusion is that the process is doomed to eventual failure, is doomed not merely to cease, but to give way to a process in the opposite direction, by which what ought to be will be displaced by what ought not, by which ethical nature will succumb to cosmic nature.

Now, if there be no God, or if being Unknowable He must be eliminated from our words, thoughts, and deeds as a negligible and useless quantity for rational purposes, it is a natural enough conclusion that right must eventually succumb to wrong. It is but a reassertion of the familiar thesis that without religion morality cannot permanently be maintained. On this occasion, however, the thesis is advanced not as a piece of religious prejudice or theological insolence, but as the teaching of science and the inevitable outcome of evolution.

FOOTNOTES:

[1] HUXLEY, Evolution and Ethics, p. 44.

[2] HUXLEY, Evolution and Ethics, p. 44.

[3] Evolution and Ethics, pp. 31, 45, 83.

IV.

IDEALISM

The bitterness of Pessimism, or rather of the pessimistic interpretation of evolution sketched in our last chapter, lies in the discovery that what we value most, what we, in our best moments, prize most highly, what we hold dearest to us, is a matter of indifference to the cosmos. That there should be any power greater than that of Right, that all goodness should in the end for ever be confounded, is incredible in the same way that the greatest losses in life are incredible in the first moment of shock in spite of the undeniable facts that show them to be real. But whereas those losses are but personal, and possibly our regrets selfish, this loss is more than personal, and the regret not merely selfish. It is not merely that we personally have held a mistaken opinion, or that any self-sacrifice--miserably small and unworthy in the retrospect--that we have made has been made for a losing cause. It is that apart from our personal share in the matter, which rated at its true value is as naught, the thing is wrong; it ought not to be. Of that we are just as certain as that our past life has not been what it ought to have been, what it might have been. The past is past beyond recall, but for the future hitherto there has been hope and faith, faith that what ought to be may be, even for us, hope that it will be so. But now, in place of hope and faith, we have the scientific certainty that the future of humanity is devoted to the triumph of the thing that ought not to be. The only consolation left to us is the inextinguishable, the unconquerable conviction that right is right even though it should not prevail. To this conviction we must hold, though the heavens should fall. To it we must hold, though it bring, as bring it must, according to Mr. Huxley, sorrow and pain and the renunciation of our own happiness.

These are hard sayings. But there is a yet harder to be added to them. Even though it should involve the renunciation of our intellectual superiority to other people, we must hold to our conviction. If we are in earnest about our moral convictions, we shall reject any suggestion that they are not after all really true, even if that suggestion seems to afford the only way of escaping from the conclusion that faith in religion has the same basis in reason as faith in science.

In proclaiming our conviction that right is right, we affirm and intend to affirm that it is so not as a matter of opinion, but as a matter of fact. In the same way, an established scientific truth is not one of those matters about which reasonable persons, who are competent to judge, may reasonably hold different opinions: it is not a matter of opinion, but a matter of fact. Indeed, both kinds of truth, moral truths and scientific truths, are quite independent of individual and personal opinion. There are people in whose opinion the earth is flat; but the earth is not flat, nor can their opinion alter the fact. There was a time when all the laws of nature were unknown to man or misconceived by him; but they operated as usual, quite unaffected by his ideas. So there are people who consider successful roguery ideal, and who would make a fortune by promoting fraudulent companies, if they could; but honesty remains a duty, in spite of their ideas. Right is right, even though there be brutes in human form; and right was right, even when the ape and tiger ruled in man, and even though they were fine fellows, in their own opinion. Cruelty and selfishness never were right at any time, and never will be. The laws of morality, like the laws of science, are objectively true: they do not vary with the opinions men entertain about them; the earth, for instance, did not move or cease to move round the sun according as men imagined Galileo to be right or wrong, nor has right ceased to be right even when the world has been most depraved.

A moral judgment, then, like a scientific judgment, is objective, not subjective; it is not the expression of a mere opinion, but the statement of a fact which has an existence independent of man. If now we ask what sort of an existence it has, it is clear that what is and what ought to be have not in all cases the same kind of existence: the thing which is may sometimes also be the thing that ought to be, but often it is not. Now, when the latter is the case, when a thing is felt to be a crying evil, a foul injustice that calls for remedy, in what sense does the justice exist on which we call to drive out the injustice? The thing which ought not to be exists, and is in possession. The thing which ought to be delays its coming. Shall we say, then, that, while that is so, it exists indeed, but exists as an ideal, as something which we know ought to be and are resolved shall be? That it must present itself to some mind or other as an object of desire, and as a possibility capable of fulfilment, is certain. That it does so present itself to man is what we mean when we attribute to him the power of moral judgment and moral action. But when we speak of man's moral judgments as being objectively true, we imply that they exist not merely in his mind, but also elsewhere. But ideals can only exist in a mind; judgments can be pronounced only by a judge. When, therefore, we affirm that in objectivity and validity our moral judgments are on a par with our scientific judgments, and that our knowledge of what ought to be is as real and true as our knowledge of what is, that the existence of ethical nature, with its demands upon our reason, is a fact as indisputable as the existence of cosmic nature, we are implicitly affirming also the existence of a mind, other than human, from whose moral judgments the laws of morality derive their validity; and as those laws are eternal and immutable, as right is right always and from eternity to eternity, so must be the mind in which they are and from which they proceed.

To say that the ideal is real sounds paradoxical. It seems like saying that to have the idea of a shilling is the same thing as possessing a shilling. That is a patent absurdity, but no one will maintain that it is an absurdity to say that we ought to try to be better than we are. On the contrary, everyone will admit that it is a truth, and a truth of the highest importance, of greater value and greater significance for our highest interests than, say, the law of gravitation, or any statement as to the ways in which matter and motion are redistributed. When the desire to amend our life is strong upon us, when we are most conscious of the heavy difference between actual amendment and amendment in idea alone, then we are most certain of the reality of the moral ideal as a fact, both of immediate consciousness at the moment and of permanent significance for us and for all men. To say that our moral convictions correspond to no real facts is simply to deny to them any validity at all. To say that the facts to which they correspond are real, but are purely subjective, being but moods, and often passing moods, of the individual, is merely to say that our moral convictions are illusions and right-doing only fancy. Nor do we mend matters if we add that all men are more or less subject to these moods, that right and wrong are purely human institutions; for if their value in the individual is naught, their existence in the multitude does but add to ciphers ciphers. On the other hand, if the moral ideal is no figment of man's imagination, if its existence does not come and go with his fitful moral struggles, then its permanent abode, the centre from which it manifests itself, must be in some permanent intelligence at the centre of things.

The Pessimistic interpretation of evolution suggests another way of reaching the same conclusion. That form of Pessimism represents cosmic nature as indifferent, if not hostile, to ethical nature; the former by its law of the struggle for existence favours the survival of the strongest and the most selfish; the latter with its moral laws strives to suspend the struggle for existence, and to defeat the selfishness which the former seeks to perpetuate and extend. Human evolution is in its essence the struggle of man as a moral being against nature as non-moral or anti-moral; and the curve traced by human evolution is the resultant of the opposition of the two forces--the microcosm, man, and the macrocosm, nature. During the first part of its course the line of human evolution rises, but during the latter part it is doomed to fall; and the curve will be completed when man, having traversed every stage of moral degradation, is merged once more in the brute matter to which originally he owed his being. Against this victory of cosmic nature man, as a moral being, protests and fights. He protests that it is wrong--wrong, not because it brings him more pain than pleasure, for right-doing also may have that result, but wrong without regard to his feelings, so that any impartial spectator who witnessed the struggle would condemn and regret the issue. If this is not so, if the condemnation is the expression merely of human prejudice, then there is nothing in the defeat of ethical nature or in the victory of its enemy, cosmic nature, really to regret; the difference between right and wrong is not an absolute or real distinction, corresponding to real facts, and the victory of cosmic nature, even if it runs counter to man's prejudices, is not thereby shown to be really wrong, though man naturally is under the illusion that it is.

The coarse and immoral piece of vulgarity which condones an act of wrong-doing on the ground that "it will be all the same a hundred years hence," is, with an extension of time, as applicable to the race generally as to the individual in particular. In a million, or a billion, years hence it will, according to the pessimistic interpretation of evolution, be all the same: matter and motion will alone exist, completely indifferent to right and wrong. What does it matter, then, whether we do right or wrong? Ultimately, it will make no difference: the distinction between right and wrong is not one of permanent value, or based on any lasting difference in things. Nor is it strange that a cause which is based on an illusion should be doomed to defeat. What is strange is that anyone should invite us to renounce happiness for such an unmeaning struggle.

The only reply to such loose talk is that it does matter, here and now, always and to all time, that right should triumph over wrong. It will not do to say that it matters now, but will not matter hereafter, for, if it is of no importance then, neither is it of any importance now. But if right-doing is the most important thing in the world, more important than happiness, more important to all time even than the perpetual redistribution of matter and motion, to whom is it important? Not exclusively, nor even primarily, to ourselves; for the essence of right-doing is the attempt to put self away and forget it, the yearning to be lifted above personal considerations and thought of self, the conviction that whilst it matters all the world to me, to do the right, the matter does not end with me. The matter is not of merely personal importance to me, nor important simply because I choose to think it so. Its value and significance are apprehended--alas! too rarely--by me, they are not created by me. Its significance and importance are real, not fictitious; that reality is not created by man, it is not a human prejudice, but exists independent of man and what he thinks. To matter and motion, those perpetual manifestations of the Power or Reality which underlies them, nothing can have any meaning or importance: it is only to a mind that things can be significant or important. If, then, the importance of right-doing is real, it is because it really matters to the Power, which underlies all things, that we should do right; and that Power must be of the nature of an intelligence, for it is only a mind which can either apprehend values or assign them. If the microcosm, man, can pass a valid sentence of condemnation upon the macrocosm, nature, it is only because and so far as his moral nature places him in direct communication with the heart of things and gives him knowledge of the will of that Power on which microcosm and macrocosm alike depend for their existence. If the distinction between right and wrong is one by which man can correctly judge between himself and the cosmos, the distinction and the judgment must proceed from a source superior to both. If it is not, then the Pessimistic interpretation of evolution falls to the ground, because it is based on the assumption that its condemnation of cosmic nature is a correct judgment. Not only does Pessimism fall, but the element of truth and reality which Pessimism contains must also be abandoned; if the distinction between right and wrong is not sufficient for the task put upon it by Pessimism, neither is it sufficient for us to build our lives on. In fine, either the ultimate defeat of the ethical process matters, or it does not. If it does not, why suffer sorrow and pain in the vain endeavour to stave it off? If it does, then to whom? No longer to man, for he will have joined the extinct fauna. Therefore to some moral intelligence to whom the triumph of right is a matter of importance.

From this dilemma the only escape seems to be frankly to admit that a billion years hence it will be "all the same," but to deny that, because it will be all the same then, it is a matter of indifference now. This argument, then, maintains that it will be all the same ultimately, and that it is an illusion to imagine that when man is extinct it can possibly matter. Here and now, however, and indeed as long as mankind continues to exist, right-doing is of the highest conceivable importance to man, more important even than happiness. But it is only as long as mankind continues to exist that it can continue to be important: its importance only exists in man's mind, and perishes with it. To say, therefore, that the ultimate defeat of the ethical process will, when established, be regrettable, is only to say that if we, or any other moral judge, were there to see it, we should feel regret about it; but we cannot possibly maintain that, because a moral judge would regret it, if he were there, therefore there will be one there to regret it. Of course, it is possible that the Unknowable may be a moral intelligence of this kind, because everything is possible with regard to the Unknowable. But we can neither affirm nor deny that or anything else about the Unknowable, for then the Unknowable would cease so far to be unknowable.

The contention of this argument, then, is that for us men, and (as far as we have any positive knowledge) for us men alone, the laws of morality are real, intensely real; but their reality begins with man and ends with man. To this contention the reply is that as regards their reality the laws of morality are on exactly the same footing as the laws of science. Take the theory of evolution for instance: from scientific observations of what is going on now it infers what has been and what will be, it reconstructs the past and forecasts the future; it frames pictures of the globe as it was before man was evolved; it forms conceptions of the earth as it will be long after man is extinct. These conceptions and pictures, however, exist only in the mind of man, for the future does not yet exist and the past has ceased to be. That is to say, evolution is an inference, or rather a mass of inferences, which like all inferences exist in the mind, could not have existed before the mind, and cannot exist when the mind has ceased to be. Science, being the work of the mind (for we cannot say that it requires no intelligence), is just man's notion of what has been, is, and will be, in the same way that morality is man's conception of what ought to be. If we say that what ought to be will cease and become meaningless when man is extinct, then we must say the same of what has been, is, and will be. If the good, the noble, the right, are merely human ideas of what ought to be, matter and motion are merely human conceptions of what is. If the reality of the former is only to be found in the human mind, so is the reality of the latter. If the reality of the one is to cease with human existence, so must the reality of the other. On the other hand, if either is to exist when mankind is no more, it is only in some mind that it can exist. It is only for a person that anything can be right or good. It is only a person that can see the past summed up or the future contained in the present. If it is legitimate and logical to infer that what is will continue to be after man's disappearance from the earth, so it is to draw the same inference with regard to what ought to be. If science is true really, and does not merely appear so to man, it must be true for some mind other than human: by an intelligence alone can truth be apprehended, or the right approved. But if truth is limited to the human mind, and ceases with it, then evolution must cease to be true when men cease to exist. Nay, in that case it cannot claim to be true at all, or rather does not claim to be true, but only seeming. On the other hand, if the law of gravity, for instance, was true before man's appearance, its truth must have dwelt in some mind. If it was not true then, we have no better reason for believing it to be true now. In fine, truth and right, what is and what ought to be, must either be dismissed as mere human imaginings, or be accepted as everlasting facts of an Eternal Moral Consciousness.

Shall we, then, say that the description which science gives of the constitution and working of the universe is indeed consistent and coherent enough with itself, and is a logical deduction from its premises, but to assert that it expresses or even corresponds to any reality beyond itself is a statement which we have no right to make? To take up this position is simply to maintain that science is consistent and logical, but that we have no reason or right to believe that it is true. If our accounts are based on imaginary figures, they may be kept as strictly as you please, but they will never show us our true position. Indeed, if our premises are incorrect to start with, the more logical our inferences are, the more certain our conclusions are to be wrong.

Shall we, then, say that the account which science gives of the cosmic process is not only consistent and logical, but expresses or corresponds to a reality? Then in that case the cosmic process, so far as it is truly expressed by science, is a logical process. But it is only a mind which can be logical or can go through a logical process. Once more, therefore, the facts of science as much as the facts of morality imply that the real is an Intelligence. In fine, the truth of science and the truth of morality are bound up together and have the same basis. If the one is valid for facts beyond the range of human observation, so is the other. If the one implies a consciousness other than human, equally so does the other.

It may be said that to regard the ruling principle of the cosmos as a moral agent is to commit the anthropomorphic fallacy. What, then, shall we say of science, which is engaged in demonstrating that the cosmic process is always logical? That science simply describes the facts as they are, and that if they are logical, it is not her fault? Then the presence of an intelligence other than human is revealed to science in the facts; and it is false to say that science merely imports her own intelligence into them. In the same way, the presence of a moral personality other than our own is revealed to us in the facts of conscience, and not imported into them by us. The presence of the Comforter is one of the facts apprehended by the religious consciousness; it is not merely the religious man's way of interpreting some other fact. From this conclusion the only way of escape is to say that anthropomorphism is a fallacy, and that it is a fallacy to which the human mind, by its very constitution, is always and inevitably subject. This argument gets rid at one blow of all indications of any intelligence or morality other than human. But how? Simply by begging the question, by tacitly taking it for granted that there is no other personality than human personality. In that case it is obvious, indeed, that man's perpetual discovery of personal power in the forces of nature, of more than human wisdom in nature's laws, and of more than human goodness in the human heart, is and must be fallacious. But only on the assumption that there is no wisdom in the world but man's, no love in all the universe but his, can we say that man reads into the facts a wisdom and a love which are not there. Are we, then, prepared to say that, in giving us a logical account of the cosmic process, science has--naturally and necessarily indeed, but none the less completely-- been mistaken? If the scientific account corresponds to the facts, then the facts behave logically. If it is the anthropomorphic fallacy to imagine that things can behave logically, then science's description of the facts must be fallacious.

Perhaps it will be sought to save science by saying that science is anthropomorphic, but not fallacious. This, however, gives away the whole case: it is an admission that, in interpreting the cosmic process as a logical process, science is simply recognising, and rightly recognising, the logical character of the facts--the anthropomorphic interpretation happens to be right. But we must note that it is not right because it is anthropomorphic, but anthropomorphic because it faithfully describes the facts. Science aims at describing and formulating facts as they are: if the laws of science are rational, it is because she found reason already in the facts, and not because she put it there. She does not make the laws of nature, neither does she dictate the behaviour of facts, nor is their behaviour merely her way of interpreting facts. Man discovers in nature wisdom, which is an attribute of personality, not because he cannot help being anthropomorphic in his views, but because nature is a manifestation of the power and wisdom of a personality other and greater than man's. So far, then, is this discovery from creating a presumption that man makes nature after his own image, that it constitutes a proof that man is made in the image of that personal will and wisdom which is expressed in nature as well as manifested in man. In discovering personality we discover what is fundamentally real in nature, and for us what is fundamentally real is also our highest ideal.

We have already remarked that paradoxical though it sounds to say that the ideal is real, the seeming paradox does express a fact--the fact at once of our consciousness of the difference between what is and what ought to be, and of our conviction that what ought to be is no mere illusion. Truth and goodness, wisdom and love, are all at the same time ideal and real. The truth to which it is the ideal of science to approximate is no mere chimera. So far as it is truth, it is not merely man's way of looking at the facts or an interpretation which he puts upon them: it is a statement of the facts, as accurate and precise as science can make it. The ideal, being an ideal, will never be fully attained; but that the truth is there to be found out is proved every time science reaches a new truth, that is to say, a truth which before its discovery was indeed not apprehended by man, but certainly was not therefore either untrue or non-existent. The truth was in the facts; for what man knows of nature he has learnt from nature, and what he finds there is not the projection of human wisdom but the revelation of a more than human wisdom. Man's knowledge is real in proportion as it approaches the ideal. The ideal is not man's surmise, or vague conception, or anticipation of what he may hereafter come to know, for such surmises are always proved to be more or less erroneous. Neither is it man's conviction that the truth exists, if only he could find it out. It is actual truth and knowledge which now exist, and, being truth and knowledge, must exist in some mind, and certainly do not exist in man's mind. Science, so far as it has approached the ideal, has done so not by being anthropomorphic, but by ceasing to be anthropomorphic--that is to say, by casting aside presumptions of what according to man's notions ought to be or a priori must be, and substituting for such preconceptions a patient, reverent study of the facts as they are.

To regard the knowledge thus gained as being at once purely human and the only reality is to say that the evolution of the universe exists only in the speculation of human thinkers, and consequently that the world as it was before man existed was created by the speculation of minds which were not in existence then and which were only subsequently evolved. How can man have been evolved out of his own speculations? How can his speculations have existed before he did? Man owes his origin to the same Power whose wisdom is revealed in nature to science, and manifested to all in all around us. Of the existence of a Power, not ourselves, we have evidence in everything that affects us. It is a fact of consciousness, but it is a fact which from its very nature does not exist solely in our consciousness. Therein it resembles ideal wisdom or goodness, which exists in us, so far as our wisdom or goodness is real, but is far from being exhausted by its partial presence in us.

The ideal in morality, again, is not the mere desire to do good or to be good, just as the ideal in knowledge is not the mere desire to know the truth. And if goodness is the object of moral desire, as truth is the object of intellectual desire, in neither case is the object of desire purely imaginary, a mere idea or conception of something which might be, but as a matter of fact is not. We do not desire imaginary pleasures or imaginary goodness, we want the reality; and to tell us that that reality exists only in idea, only in our own imagination, is a misleading half-truth. True, we must have some idea of it, or else we could not desire it. But neither could we desire it if it were presented to us as purely imaginary. In other words, the object of moral desire is apprehended, at the moment of apprehension, as both actual and possible, as existing simultaneously for us and beyond us. The case is the same with ideal truth: we could not desire it, unless we had some conception of it, unless it were to some degree or in some way present to our consciousness; yet, at the same time, the knowledge which we desire to have but do not yet possess is certainly, so far as we do not possess it, beyond our consciousness. It is because we have not got it that we want it. And the object of desire, what we want, is not imaginary truth, but real truth; just as in our better moments we want to do not what we imagine to be right, but what is really right. The Real, therefore--real truth, real goodness--is apprehended, at the moment of apprehension, and desired, at the moment of desire, as existing both for us and beyond us.

The proviso, "at the moment of apprehension, at the moment of desire," is important, because it strikes at the root of all forms of subjective idealism. They all assume that the only thing directly apprehended is what exists for us; that consequently the supposed existence of any real thing or person beyond us is a mere inference, and an inference the truth of which we have no means of checking, because it is a statement about things of which we have no direct apprehension or knowledge. On this assumption, therefore, the only things man directly apprehends are his own states of consciousness, his own sensations, etc. Are we to call them real or not? If they are not real, his whole life is a dream, his speculations fancies, and his desires illusions. If they are the only reality of which he can be certain, then the only truth is that which man knows, the only good is that which man does, the only world is that which man thinks, the only God is that which man makes, the magnified, non-natural shadow of man projected on to the mists of the Unknowable.

It is important, therefore, to insist that the Real--the reality of existence, of knowledge, of goodness--is not an inference, but a matter of direct apprehension. It is certain that goodness or knowledge to be an object of desire must be presented to us in idea; but it is equally certain that the mere idea is not what we desire. The object of desire is directly apprehended as in our consciousness and beyond it. The natural world around us is also directly apprehended as at once in our consciousness and beyond it: it is presented to our minds, but it is presented as real.

It is important also to note that the real does not forfeit its reality to our apprehension when and because it takes up its abode in us: goodness does not cease to be good because we do it, nor truth cease to be truth because we know it. It does not follow that because the ideal cannot be fully realised, it cannot be realised at all. On the contrary, the conviction that it cannot be completely attained is itself the guarantee that it can be attained partially. Yet it has been assumed that if a thing is apprehended by us it cannot be real, that real knowledge begins just where our knowledge ends, that the further we push our knowledge forward the further real knowledge recedes from our view. On this assumption is built the theory of the Unknowable, the theory that whatever is known to man is a state of man's consciousness; that states of consciousness are subjective, are merely the appearances of things, not the things themselves; that the real things, the things themselves, are unknowable; their appearances alone can be known to man; therefore the real is for ever unknowable. "The reality existing behind all appearances is, and must ever be, unknown."[4] Consequently, inferences about the Real are valueless and futile. By way of compensation, however, our knowledge of the unreal is, on this theory, varied and extensive: it includes, for instance, the theory of evolution and the whole of science.

But the assumption which leads to this strange conclusion is opposed to the facts. The fact, as we have contended, is that the real in consciousness is continuous with the real beyond consciousness, and is apprehended, at the moment of apprehension, as being thus continuous, and is not reached by any process of inference. The real is not a matter of inference, but of apprehension. Its existence cannot be deduced from anything else; it is that from which all conclusions must be deduced. I cannot prove that a thing is real any more than I can prove that I have toothache. There is no need.

FOOTNOTES:

[4] HERBERT SPENCER, First Principles, ch. iv. ?22, p. 69.

V.

THE REAL

We began, at the beginning of this book, by accepting Evolution as a fact, as all ordinarily educated persons in the present state of scientific knowledge are practically bound to do. Accepting it as a fact, we proceeded to inquire what, if anything, it had to tell us about the moral government of the world; and we found that very different interpretations were put upon the theory of Evolution by different authorities. According to one interpretation the process of Evolution was a continual progress from good to better: good could only give way to higher good. According to another interpretation goodness was a transient, evanescent phase in the process of evolution, of no permanent value: the ethical process was doomed to be defeated by its enemy, the cosmic process. According to a third interpretation the notion of good was a pure illusion, necessary indeed, inasmuch as without it there would be no survival for man in the struggle for existence, but none the less an illusion.

Much as these interpretations differ from one another as to the moral significance of the process of evolution, or indeed as to whether evolution has any moral significance at all, they are agreed upon one point. They are agreed that it is impossible to draw any inference from the facts of evolution as to the moral government of the universe. To affirm its moral government would be to claim knowledge of the Unknowable, which is an obvious absurdity. It would be to attribute power, consciousness, wisdom, and goodness to the Real; and the Real is and must ever be unknown.

This identification of the Real with the Unknowable leads us into the following ridiculous impasse: the vast majority of men look, and must always look, for guidance and information to science and theology; and theology is knowledge of the unknowable; science, knowledge of the unreal. Those who are content with this blind alley may remain in it. We propose to try to find our way out of it.

If we analyse our perception of any material object, that is to say, of any object which we perceive by means of the senses, we shall find that our perception of the object consists of the sensations which we have of it. To perceive an orange is to see that it is yellow, to feel that it is round, to smell it, taste it, and so on. These various sensations together constitute our perception of the orange. Now, the subjective idealist says that the perception is the orange, and that the orange is the perception. To the beginner in philosophy that sounds absurd: he knows that his perception is not the orange, and that the orange is something more than his perception of it. But when he is asked, "What more? If the orange is not the perception, what is it?" he does not generally produce any satisfactory reply; and then he is told that his notion, that there is anything in the orange except his own perception or sensations, is obviously not a fact of sensation or a thing directly observed, but merely a belief or inference of his. On the other hand, he generally puts a very natural question to his instructor: "If the orange is merely my perception, what becomes of the orange when I do not perceive it? Granted that it exists whenever I look at it, what becomes of it in the intervals when I am not looking at it? Does it exist then, or does it not?"

To this Bishop Berkeley replies that it does; that it exists then in exactly the same way as it does now, that is to say, it exists in idea (i.e. perception or sensation); but as it does not exist in my perception, when I am not looking at it, it must exist in the perception of some other mind, to which all things at all times are present.

With the fact which forms Berkeley's conclusion I have no quarrel. What I should like to show is that it does not follow from these premises.

Berkeley's argument is: All men believe, and rightly believe, that the things they see are permanent. The things they see are ideas (perceptions, sensations) of a mind. Therefore the permanent world is the idea of a permanent mind.[5]

But "the things they see" is an ambiguous expression. If by "the things that I see" is meant "my sensations of sight," then they are not permanent, for they only last as long as I look at the object, and consequently any argument based on their supposed permanence falls to the ground. On the other hand, if "the things that I see" are permanent, then they are not merely my sensations of sight--in which case subjective idealism is wrong, and my perception of a thing is not the whole account of the thing and does not exhaust its reality. The things which I perceive are not my sensations: they are things of which I have sensations. In fine, they are apprehended, at the moment of apprehension, as being both within and without consciousness.

To the question whether a thing exists when I am not looking at it, John Stuart Mill replies, in effect, that as often as I look at it I shall see it; that if I were looking I should see it. This is true enough; but it is no answer to the question. When further pressed, Mill further replies that, if things do not exist when we do not look, we should nevertheless necessarily be deluded by the association of ideas into imagining that they do exist when not looked at. Here, again, it is perfectly true that, if things are not real, it is a delusion to imagine they are. But that is no answer to the question. It is, in fact, a question which the subjective idealist cannot answer. To say "No! Things out of consciousness are non-existent," is to say that effects of which the causes are unobserved are effects produced by non-existent causes. To say "Yes" is to admit that things can exist out of consciousness as well as in, which is what subjective idealism is there to deny.

We submit, then, that the analysis of experience which subjective idealism makes is not an exhaustive analysis; and that, when the man of common sense says that in looking at anything he is aware both of his sensations of sight and of something more, he is stating the actual facts as they are given in experience to all of us.

We apprehend a thing as being both our sensations and something more. When the idealist says that the latter half of this apprehension is a misapprehension, he rejects an observed fact of experience, not because he does not find it in his experience, but because it seems to him impossible that it should be there. He argues that to say we can be conscious of what is not in our consciousness is to say that we can be conscious of something of which we are unconscious--a patent nonsense. He might admit, for the sake of argument, that possibly a thing could exist both in consciousness and out, and even that we might know that it so existed. But he cannot admit that a man is conscious of what he is not conscious of.

He is not required to admit it. He is required to admit that our perceptions are not the only things of which we are conscious; or, to put it in other words, that our states of consciousness are not the only things of which we are conscious. And he is required to admit it simply and solely on the ground that it is a fact of common observation and everyday experience. Thus, for instance, we perform actions, and (usually) we are conscious of performing them. But the action is something more and other than our consciousness of it. Or is someone going to maintain that doing and knowing are the same thing? Is anyone prepared to push the illusion-argument so far as to say that the idea that we do things is a mere delusion? If it is not a delusion, if it is, on the contrary, a fact, then our actions are not states of consciousness, but things of which we are conscious. We apprehend them, in the very act of apprehension, as realities distinct from the consciousness which we have of them. And we have the very same guarantee for their reality as we have for the reality of our perception or sensations of them, viz. the fact that we are conscious of them.

In the same way, when we push a solid object or feel the impact of a moving body, we are as conscious of that body as of our muscular sensations: our sensations make up our perception of the object, but are not the object. They constitute the state of consciousness, but that state is not the only thing we are conscious of. The object is apprehended as being in consciousness and not as merely being our consciousness of it.

Mr. Herbert Spencer, at least, is quite clear that our states of consciousness are not the only things of which we are conscious; he holds even that we are vaguely conscious of that which transcends our consciousness. Thus, our personality is not a state of consciousness, yet we are conscious of it, and "its existence is to each a fact above all others the most certain."[6] And, as for the real, "our firm belief in objective reality, a belief which metaphysical criticism cannot shake," is not merely "a positive though vague consciousness of that which transcends consciousness," but "has the highest validity of any"[7] of our beliefs.

But though Mr. Spencer admits, or rather insists, that we know that the Real is, he denies that we know what it is. In other words, he accepts the validity of one half of every act of experience and denies the validity of the other half. Our analysis of experience has shown us that we apprehend the real, in the very act of apprehension, as being both a state of our consciousness and something more than that state. To say that one half of the apprehension is a misapprehension is to say that both are invalid. If what is present in consciousness is merely appearance and not the real thing, then our states are the only things of which we are conscious, and the existence of anything more is not a fact of experience and observation--still less can it have the highest validity of any of our beliefs.

We may be asked, "Granted that the Real is more than a state of our consciousness, what more is it?" and, if no answer is forthcoming, we may be told that after all then, it seems, we know that the Real is, but not what it is. The reply is: So far as the Real is out of consciousness we may not know what it is; as far as it is in, we do. By "being conscious of a thing" we mean knowing what the thing is--not necessarily complete knowledge, but some.

If it be said that, on our own showing, a thing and the knowledge of it are different, and that consequently however great our knowledge may become there always remains, and must always remain, something which we cannot know, because it is ex hypothesi, not knowledge, we must reply that this objection is but a restatement of the inveterate fallacy of idealism--the fallacy that states of consciousness are the only things we can be conscious of; that if we know a thing the thing ceases to be anything but our knowledge of it; that to be conscious of performing an action is proof that no action is really performed, and that the only doing is knowing.

We act, and we know that we act. Reality must be accorded to both or denied to both; it cannot be accorded to one and denied to the other. Indeed, knowledge itself is action, a series of actions. But it is also something more, just as an action of which we are conscious is something more than our consciousness of it.

But we are conscious not only of our own actions, but of the reactions of things on us, and of the interactions of things on one another. We apprehend all three--action, reaction, and interaction--as real; we know not only that they are, as being realities, but also what they are as states of consciousness. As states of consciousness they are successive sensations or perceptions; as more than states of consciousness they are power or force.

The study which science makes of the interactions of things on one another reveals those interactions as conformable to law and happening in such a way that their occurrence can be logically deduced, and even foretold, from their laws. In a word, they happen in a way that can be reasoned out, and they constitute together a logical process. The reality, the power, the activity which is exhibited in this process is exhibited therefore as a rational activity, as reason active; and both the reason and the action are apprehended by us as real, and not as mere states of our consciousness.

If the scientific account of the universe and the theory of evolution, so far as they are true, are not mere exercises of the imagination, but represent events and changes which actually have taken place and are taking place beyond the range of actual observation, it must be because they are logical inferences from real events and real changes which are matters of direct observation. If the observed events have no reality, we have no ground for believing the unobserved or inferred facts to have any. Unless the real events follow a logical sequence, our inferences must be fallacious in proportion as they are logical. We believe the inferred facts to be real because we believe the observed facts to be real; and the observed facts are presented to us and apprehended by us to be not merely our sensations but also realities. On no other ground can we or do we trust science to guide us in life.

Nor do we trust morality on any other ground. So far as we trust the impulse to do right, or base any calculations upon it or draw any inferences from it, we do so because we apprehend it, in the act of apprehension, as both a state of our consciousness and something more. As in the impact of a moving body we apprehend not merely our sensations, but also the presence of a real power, so in the impulse to good we apprehend not only our consciousness thereof, but the presence of a real power, with regard to which we know not only that it is, but to some extent what it is--a power which would have us do good and be good.

If material things are but ideas of ours, so the Right and Good may be. If the latter are mere aspirations and nothing more, the former are mere sensations and nothing more. But if in things we are conscious of a power not ourselves, so are we in our consciousness of the Right and Good: our aspirations are inspirations. We apprehend their reality in exactly the same way as we apprehend the reality of material things--by direct observation. And we have exactly the same evidence--the evidence of immediate consciousness.

"Let no man spoil you with philosophy." The statements that "knowledge is the only reality," "the only Real is the Unknowable," are contradictory not only of each other, but of those facts in the common experience of mankind which afford the only safe foundation for philosophy as well as for science. Both statements logically imply that our only knowledge is of the unreal; and from knowledge of the unreal to the unreality of knowledge is a necessary step. But existence is not merely knowledge: existence is also action. A thing is that which it does, and not merely that which it is known to do. Or rather a thing never does anything: only a person can act. The "action" or "behaviour" of a thing is only a metaphor.

FOOTNOTES:

[5] See Appendix on Bishop Berkeley's Idealism.

[6] First Principles, ch. iii. ?20.

[7] Ibid., ?26.

VI.

EVOLUTION AS THE REDISTRIBUTION OF MATTER AND MOTION

Assuming the process of evolution to be a fact, we have inquired what is the value of that fact, what significance it has for man as a moral being, anxious to direct his life in accordance with the best lights he can obtain. In our attempts to draw any inference from the facts of evolution as to the moral government of the universe, we have always found ourselves ultimately confronted by the notice--The Real is Unknowable. Obviously, if "the ultimate of ultimates," the Real Power or Force, of which all things and beings are manifestations, is unknowable, we cannot know whether it cares or does not care for what is true or good. But if the Real is Unknowable, then the knowledge which we do possess is not knowledge of the real, and consequently all our science is unreal knowledge; the theory of evolution is a system of delusive inferences from unreal facts. That, however, is a thing which we could not believe. Doubtless our knowledge is small compared with our ignorance. Doubtless there is much which the human mind could not understand without becoming more than human. Doubtless, also, every addition to our knowledge involves a readjustment and correction of our previous inferences; and a considerable addition, such as the theory of evolution was, causes a considerable change in our conception of the universe and its laws. But all these admissions cannot compel us to admit that science is wholly unreal knowledge, or that evolution is an entirely unreal process. We sought, accordingly, to show that we have some, if only partial, knowledge of the real, that that knowledge is not wholly inferential, but that so far as it is inferred it is inferred from real facts, the reality of which is directly apprehended in the common experience of mankind.

As a matter of fact, those writers who proclaim the unknowability of the Real, when they are writing as philosophers, abandon it when they are engaged in science. When they are working out the theory of evolution, they take it for granted that the process of evolution is a reality, that the common experience of mankind is trustworthy to some extent, and that to that extent the Real is knowable and known. They assure us that, though the knowledge we have is not knowledge of the Real, it is just the same for us as if it were--if the Real could enter into our consciousness, we really should not know the difference. "Thus then we may resume, with entire confidence, those realistic conceptions which philosophy at first sight seems to dissipate."[8]

On examination, however, it turns out that the entire confidence which is thus restored to the reality of material things is not extended to the reality of those ideals of the good, the beautiful, and the holy which play their part in the lives of men and in the evolution of mankind--or not to all of those ideals.

Now, it is scarcely to be hoped that a theory which begins by ignoring certain facts in the common experience of mankind, or by denying their reality, can end in a satisfactory explanation of them. Either it will be consistent and proclaim them to be illusions, or it will be inconsistent and quietly include them from time to time as it goes on--in which case the explanation it gives of them will be no explanation. Thus, for instance, as we have already argued, the Optimistic interpretation of evolution, professing to exhibit the Ideal of morality as one of the ultimate consequences of the redistribution of matter and motion, ends by denying any difference between what is and what ought to be, and thus reduces the moral ideal to a mere illusion. The Pessimist, on the other hand, insisting on the reality, and to some extent the supremacy of the moral ideal, confesses his inability to explain its validity as being due to evolution: the fact that it has been evolved does not account for its validity, because the tendency to evil has been also evolved, but is not, therefore, to be yielded to.

The object of this chapter is to examine the hypothesis that the process of evolution is nothing but a perpetual redistribution of matter and motion, and to show that the hypothesis cannot explain, and as a matter of fact does not explain, all the facts which it is framed to account for.

The theory of evolution is an attempt--one of many attempts that men have made--to explain the process by which the totality of things has come to be what it is. It differs from most other attempts in that it endeavours to give a scientific explanation of the process, and that consequently it does not profess to go back to the beginning or to discover the origin of the process.

The nature of scientific "explanation" is well understood by men of science (in England, at least), and has been made familiar to the non-scientific world by John Stuart Mill. An event is scientifically "explained" when it is shown to be a case of a general law; a law is "explained" when it is shown to come under some more general law. In other words, the business of science is to show that the thing under examination always happens (or tends to happen) under certain circumstances which science can formulate with more or less exactness. But how or why the thing should happen thus, science does not undertake to explain: "what is called explaining one law of nature by another, is but substituting one mystery for another; and does nothing to render the general course of nature other than mysterious: we can no more assign a why for the more extensive laws than for the partial ones."[9] It is only "minds not habituated to accurate thinking" which imagine that the laws are the causes of the events which happen in accordance with them, "that the law of general gravitation, for example, causes the fall of bodies to the earth."[10] It may be a law of science, a perfectly true statement, that the phenomenon B always follows the phenomenon A; but that statement, true as it is, is not the cause of B. That A is always followed by B is demonstrated by science. Why it should be followed by B is as mysterious as magic--as mysterious as that the waving of the magician's wand should be immediately followed by the rising of a palace from the ground. How the one thing can follow the other, is no part of science's business to explain.

Science, therefore, is essentially descriptive: with ever-increasing accuracy it describes things and the order in which they happen. Evolution, then, as a scientific theory, is also purely descriptive: it describes the way in which things have come to be what we see them to be, the process by which the totality of things has come to be what it is. But when the purely scientific and descriptive part of the work is done, when science has formulated the order of the events which have led up to the existing state of the universe, when the process of evolution has been described, there still remain the questions which science refused even to try to answer, and there also remain other questions more vital to science. There arises the question, In what sense is evolution a real process? do the laws of science exist only in the minds of men of science? is the process of evolution merely the description which is given of it (as according to some thinkers a thing is only the sensations which we have of it), or is it something more?

Obviously the question whether evolution is a real process, whether there is any reality in science, is one which cannot be answered, either in the affirmative or in the negative, without some idea of what "reality" means, of what the "real" is. "What is the meaning of the word real? This is the question which underlies every metaphysical inquiry; and the neglect of it is the remaining cause of the chronic antagonisms of metaphysicians."[11] Before we are logically entitled to say that evolution is a real process, we must answer the question, "What is the essence, the ultimate reality of things? who or what is the Being that is manifested in 'all thinking things, all objects of all thought'?"[12]

Now, to these questions, as to the Being and Becoming of the universe, science has nothing to say. Science does not even afford the materials for an answer to them, any more than to those other questions as to how or why things should happen in the way which science describes. Science describes things, but does not undertake to prove that things exist. Science is organised common sense, and common sense takes it for granted that things exist. Having made this assumption, science proceeds to investigate with scientific exactness the order in which events succeed one another and co-exist with one another, within the range of direct observation; and infers that, even when they are beyond the range of direct observation, they continue to occur in the same order of sequence and co-existence. But here again science refuses to have anything to do with any metaphysical questions as to how or why things should thus occur. All sorts of conjectures may be made, and have been made, to explain why B should follow A, or co-exist with it. But science is not pledged to any of them. The only thing she undertakes to show is the fact of the sequence or co-existence; and this she can do without assuming the truth of any of these conjectures. Indeed, the progress which science has made is largely due to the fact that she has steadily declined to have anything to do with such conjectures--having found out by experience that they simply distract her from her proper business of observing with the utmost exactness what actually does take place. It may be that A in some mysterious and wholly inexplicable way "produces" B, that is to say in technical phraseology, is "the efficient cause" or "mechanical cause" of B. It may be that the sequence of B upon A is a volition of the Being which is manifested in all thinking things, in all objects of all thought. Science cannot prove, and will not even discuss, either suggestion: she confines herself to the assertion that, as a matter of careful and exact observation, B does follow A. Whether we call A an efficient cause or not, matters not to science: call it so or refuse to call it so, the fact once established by science, that B follows A, remains. The theory of efficient or mechanical causes is doubtless of importance, but not to science. If it is proved to be false, not a single fact of science is shaken.

The mechanical theory may be true or may be false, but in either case it is a metaphysical theory. If science is descriptive--descriptive of the uniform succession and co-existence of facts--then science no more proves the mechanical theory to be true than it proves the volitional theory to be true. Both are theories as to why facts should succeed one another in the order described by science; and science does not undertake to prove the truth of such theories, nor does she wait for them to be proved or disproved.

Many men of science, however, are also philosophers, and hold, as they are fully entitled to hold, that the mechanical theory is the true interpretation of nature. Now, "mechanics is the science of motion; we can assign as its object: to describe completely and in the simplest manner the movements which occur in nature."[13] On the mechanical theory, therefore, "the object of all science is to reduce the phenomena of nature to forms of motion, and to describe these completely and in the simplest manner ... the only complete description is that afforded by a mathematical formula, in which the constants are supplied by observation. This permits us to calculate those features or phases of phenomena which are hidden from our observation in space or in time."[14] This, we need hardly add, is in agreement with Mr. Herbert Spencer's view of the theory of evolution as a description of the process of the redistribution of matter and motion.

It seems, then, that according to this particular metaphysical theory, which maintains the mechanical explanation of nature to be the true one, the object of all science is to describe (with mathematical accuracy, where possible) the movements of things in space. But science is universal; evolution extends to the whole cosmic process. Therefore, the only things with which science has to do, or which are factors in the cosmic process, are things moving in space.

As a metaphysical argument this theory seems to us unsatisfactory. It converts, simply and illegitimately, the proposition sanctioned by common sense, that material things are real, into the proposition opposed to common sense, that all real things are material. It assumes, apparently unconsciously and certainly without proof, that the only things capable of scientific description are movements in space, the only laws in the universe mechanical laws.

Historically, material things were the first to be studied and described with scientific exactness. It is only natural, therefore, that the methods and assumptions which have been employed with conspicuous success by the physical sciences should be extended, tentatively at least, elsewhere. It is equally natural that protests should be raised, and the extension proclaimed by philosophers to be illegitimate--"impoverishing faith without enriching knowledge."[15] "To regard the course of the world as the development of some blind force which works on according to universal laws, devoid of insight and freedom, devoid of interest in good and evil, are we to consider this unjustifiable generalisation of a concept, valid in its own sphere, as the higher truth?"[16]

It is not, however, likely that science will drop a generalisation, however "unjustifiable" in metaphysics, if it works in practice. The question is whether it does work; and that is plainly a question of fact, not a question of metaphysics. We want to know therefore, first, whether things moving in space are the only things with which we are acquainted in common experience; and, next, whether all the changes which take place within the range of scientific observation are or can be explained by the laws of mechanics.

It is clear that, if the mechanical theory of science and of evolution is to be successfully maintained, both these questions must be answered in the affirmative. It is equally clear that, if we confine ourselves to the actual facts, both questions must be answered in the negative.

Thoughts, ideas, conceptions, sensations, feelings, emotions are things of which we have experience at every moment of our waking lives; and none of them are things which occupy space or move in space. A thought is not a thing which can be measured by a foot-rule, as things in space can be; the greatness of an idea is not one which measures so many yards by so many; a conception has no cubic contents; a toothache cannot be put in a pair of scales, nor can any process of chemical analysis be applied to hope or fear. We find ourselves, therefore, in this dilemma: if the mechanical theory is true, and science can deal only with things moving in space, then psychology and sociology are not sciences, and their subject-matter never can be made amenable to scientific treatment. On the other hand, if psychology is a science, then science deals with things which do not move in space.

We submit that psychology is a science, that our sensations, emotions, ideas, etc., can be observed, and can be described scientifically, that is to say, that their uniform sequences and co-existences can be stated with accuracy and formulated as laws. We submit further that our definition of science should be based on facts, and not framed to suit a metaphysical theory. A satisfactory definition of science must include all the sciences. The definition put forward in the interests of the mechanical theory excludes arbitrarily the mental and moral sciences, and implies that their subject-matter is beyond the power of science to deal with. The exclusion and the implication are consequent upon the suggested limitation of science to things moving in space, and are of the essence of the mechanical theory. Both the exclusion and the implication are unnecessary if we adhere to the older conception of science, as it occurs in Mill, which claims for science all phenomena of which the sequences and co-existences can be observed, described, and formulated as laws.

What we have said with regard to science applies also of necessity to Evolution. If Evolution is simply the continual redistribution of matter and motion, if matter and motion are the only things subject to evolution, then consciousness and conscience are not subject to evolution. On the other hand, if they too have had and are having their evolution, then the redistribution of matter and motion does not sum up the process of evolution, and is not a correct statement of the process. If it were an induction drawn from a consideration of all the facts of evolution, it would cover them all. But it does not: it excludes a large class of important facts, because their exclusion is demanded in the interests of a particular metaphysical theory--the mechanical theory. It implies that the operation of evolution is confined to a limited set of facts. If the implication is false, then evolution is a bigger thing than the mere redistribution of matter and motion.

The way in which it is usually attempted to force the mechanical theory to square with the facts, or rather to cut the facts to fit the theory, is to point to the connection between the mind and the brain, and to proclaim the consequent dependence of mind on matter. Now, that there is a connection between mind and brain is certain. What the connection is exactly is as yet uncertain. But the fact that two things co-exist, are connected with one another and vary together, does not prove that the one thing is the other. On the contrary, it postulates that the two things, though related, are different. The mechanical theory either commits the fallacy of mistaking connected things for identical things, or it fails to prove the very thing necessary for its justification, viz. that thoughts, emotions, etc., are things occupying space and moving in space. The chemical and physiological changes which take place in the brain are movements in space. But it does not follow that the corresponding pains or ideas float about in the air or move from one point in space to another.

Further, as a metaphysical theory, this identification of matter with mind is a double-edged weapon: it cuts both ways: if mind is matter, matter is mind; if mind is thinking matter, then matter is latent thought; and thought is consequently exhibited not as being the last product of evolution, but as a factor in it from the beginning. But this identity of mind and matter is a purely metaphysical speculation: it is a conjecture to explain how it is that two phenomena can co-exist in the way in which they are observed to do. Such conjectures science does not require: she does not undertake to explain why things are, but to describe--if possible with mathematical exactness--the order of their sequence or co-existence. This function science can discharge equally well whether the changes of consciousness are or are not supposed to be movements in space. Metaphysicians may argue the point; in the meantime science is describing and formulating the laws of mind and endeavouring to correlate the changes of consciousness with the physical changes of the brain and the nervous system. The mechanical theory neither helps nor hinders science in her work.

But science does throw some difficulties in the way of the mechanical theory; or, rather, the facts of science refuse to fit into the theory. If the stream of consciousness is nothing but a series of physiological and chemical changes, the laws of the one ought to be identical with the laws of the other, and both with the laws of mechanics, on the mechanical theory. But they are not. Those concise descriptions of mental phenomena which constitute the laws of psychology ought to coincide with those other concise descriptions of fact which constitute the laws of chemistry, if the facts described by the two sciences are the same. But the two sets of laws have, to say the least, more differences than resemblances.

This brings us to our second point. Our first point was that if the concise description of evolution, which sums it up as the process by which matter is continually redistributed in space, is to be proved to be true, it must be shown that movements in space are the only events which we know to take place. Our second point is that, unless it can be shown that mechanical laws are the only laws at work in the universe, this description of evolution does not find room for the whole working of the process of evolution.

Whether the only laws in the universe are mechanical laws is primarily a question of fact; and on the facts, as known to us at present, the answer to the question is a decided negative. The laws of psychology and of ethics are neither identical with nor have they been deduced from any physical laws. As a hypothesis designed to explain the way in which the world works, the redistribution of matter and motion neither includes nor accounts for those laws which are of most importance to man.

This appeal to the facts which are actually known is, however, often conceived to be in reality an appeal to our ignorance: mental laws have not as yet been shown to be deducible from physical laws, but they may be. So, too, the fact that no attempt to extend the gravitation formula from astronomy to any other department of science has yet succeeded, is no proof that it never will be so extended. Neither, we may remark, does it constitute any presumption that it will. Are there, then, any other grounds for presuming that mental law may yet be shown to be merely a case of some physical law? To some minds there seem to be grounds for presuming that it not only may, but must. However great our ignorance of the details of the process of evolution, there are certain broad facts which are beyond dispute. It is indisputable that there was a period in the history of the earth when there was no life upon it; that the elements which constitute living matter are themselves lifeless; that consciousness is correlated somehow with those organic compounds, the elements of which are inorganic. These facts together constitute an irresistible presumption that ultimately mind and matter must obey the same laws.

But this is not the desired conclusion. The conclusion desired is that mind must obey matter's laws. The fact that mind and matter obey the same ultimate laws is a different thing, and rather indicates that even the redistribution of matter and motion requires ultimately some other explanation than merely mechanical laws afford. To the religious mind it is quite intelligible that mind and matter should obey the same laws--God's laws.

It may be said, however, that we have not done full justice to the presumption raised by the broad facts of evolution. When there was no life upon the earth, the only laws in operation must have been physical laws, and consequently the laws of life and consciousness must have been produced by the laws of matter.

Now, this argument in effect amounts to a denial of any difference between the mechanical composition and the chemical combination of bodies. Bodies when mechanically compounded continue to follow the same laws as they obey when uncompounded, and their conjoint action can be deduced and foretold from the laws to which they are subject in their separate state: "Whatever would have happened in consequence of each cause taken by itself happens when they are together, and we have only to cast up the results."[17] With chemical combination the case is quite different: the chemical compound exhibits properties and behaves in ways which are quite different from the properties and behaviour of its elements, and could not be foretold from any observation of them. Water, which is a combination of oxygen and hydrogen, exhibits no trace of the properties of either. "If this be true of chemical combinations, it is still more true of those far more complex combinations of elements which constitute organised bodies, and in which those extraordinary new uniformities arise, which are called the laws of life. All organised bodies are composed of parts similar to those composing inorganic nature, and which have even themselves existed in an inorganic state; but the phenomena of life, which result from the juxtaposition of those parts in a certain manner, bear no analogy to any of the effects which would be produced by the action of the component substance considered as mere physical agents.... The tongue, for instance, is, like all other parts of the animal frame, composed of gelatine, fibrin, and other products of the chemistry of digestion, but from no knowledge of the properties of those substances could we ever predict that it could taste, unless gelatine or fibrin could themselves taste; for no elementary fact can be in the conclusion which was not in the premises."[18]

What is thus true of physiology and of those chemical combinations on which it is based, is true also of sociology and the psychological facts on which it is based. "When physiological elements are combined, the combination reveals properties which were not appreciable in the separate elements. The increasingly complex combination or association of organic elements may produce an entirely special set of phenomena.... Their combination exhibits something more than the mere sum of their separate properties. Thus, no knowledge of man as an individual would enable us to forecast all the institutions which result from the association of men and which can only manifest themselves in social life."[19]

It is clear, then, that the mechanical theory of evolution can only maintain itself by obliterating the distinction between mechanical juxtaposition and chemical combination. The obstacles which stand in the way of this obliteration, at the outset, are two. First, the behaviour of a chemical compound bears no resemblance to the behaviour of its constituents when separate. Next, the laws of the compound cannot be deduced or exhibited as consequences of the laws of the separate elements. To these two objections it may be replied, first, that though the compound bears no resemblance to its separate constituents, the character of every aggregate must be determined by that of its component parts; and, next, that with more knowledge we shall come to see the way in which the laws of the separate components generate the law of the whole. Perhaps, by way of illustration, we may employ an analogy. A number of bricks can be placed on one another to form a cube; a number of cannon balls will form a pyramidical pile. The aggregate of bricks resembles in shape the separate bricks; the aggregate of balls does not resemble a ball in shape. Yet the pyramidical shape of the pile of cannon balls is as certainly determined by the shape of the separate balls, as the cubical shape of the heap of bricks is determined by that of the separate bricks. Now, we do not know the geometrical structure of chemical atoms; but, on this analogy, it is reasonable to suppose that, if we did, we should see at once that the structure of a chemical compound is dependent on, though different from, that of its elements. So too in sociology, the aggregate, society, is not a human being, but the character of any given society is determined by the character of its individual members.

This last illustration, however, brings us to a fresh difficulty in the way of the mechanical theory. As is observed in the remarks quoted previously from Monsieur Bernard, the peculiar characteristic of those more intimate combinations which form the subject-matter of chemistry, physiology, and sociology is that in them the combining elements reveal properties which were not perceptible in them previous to their combination. It may be true that the character of these more intimate combinations is determined by the properties of their constituents, but it is by the properties which they reveal when in combination, not by those which are manifest in them when uncombined. Therein lies the difference between mechanical compounds and chemical combinations; and it is that difference which the mechanical theory does not account for. The more intimate, chemical, physiological, and sociological combinations take place in virtue of properties which require the combination to reveal them. In sociology it is not the juxtaposition of individual men, but their co-operation, which makes a society. In chemistry, the formation of chemical compounds implies the affinity of the elements.

It seems, then, that the mechanical theory contains half the truth, but not the whole truth. The half-truth which it insists on is that in both mechanical and chemical combinations there is juxtaposition of the constituent elements. The half of the truth which it overlooks is that when the elements are juxtaposed in one way they develop or manifest new qualities, when juxtaposed in the other way they do not. The mechanical theory asserts that the only factors in evolution are matter and the force which moves matter about: it takes into account the external factors, but leaves out the internal force or spontaneity in virtue of which things in a suitable environment develop new qualities. Doubtless the juxtaposition of the elements is a condition without which they would not manifest their new properties: the redistribution of matter and motion is a condition of evolution, but it does not constitute evolution. Rather, it is the continual revelation of these new qualities which constitutes evolution, chemical affinity and all its consequences in chemistry, spontaneous variation and all its consequences in the evolution of organic life.

From this point of view it becomes clear why the laws of a chemical compound neither are nor can be exhibited as consequences of or deductions from the laws of its separate constituents. The properties which the law of the compound describes are not the properties which the separate elements exhibit. The living matter of biology, the active atoms of chemistry, are not products of the lifeless, inert matter of mechanics; but are different and higher revelations of the same power which is manifested in different degrees in all. It is this progressive manifestation, and not the mere drifting about of bits of matter, which constitutes evolution. The redistribution of matter and motion may be a concomitant of evolution, but it is not evolution. "The continuous adjustment of internal relations to external relations," which Mr. Herbert Spencer[20] offers as a definition of life, may be a condition of the maintenance of an organism; but life is and means to each one of us, and to the humblest thing that breathes, much more than that. Mr. Spencer's definition of life leaves, for instance, consciousness out, as of no account in life, and would be equally applicable to many automatic, self-adjusting machines. The definition constitutes an admission that life and consciousness cannot be exhibited as a consequence of the redistribution of matter and motion. They appear at a certain (or uncertain) point in the process of redistribution, and they have as concomitants certain further redistributions; but they are neither the consequence of nor are they identical with that redistribution; nor can their laws be reduced to mere cases of the laws of matter and motion.

The doctrine that evolution consists in nothing but movements in space, amounts to the assertion that we know nothing about things and men except that they move. In point of fact, we know a good deal more. We know that men think, and that the movement of thought is not a movement in space. We know that the vibrations of the ether are movements in space and that they are also something more: they are known to us also as sights, sounds, etc. No explanation or concise statement of the process of evolution can be satisfactory, or even scientific, which begins by denying the relevance and even the reality of the most important part of our knowledge. If the "first principles" of evolution are to be scientific, they must be inductions drawn from observation and based on some similarity in the phenomena observed, in which case, and in which case alone, they will apply to both classes of phenomena, mental and material. If any "principle" is true of one class alone, it is shown thereby not to be a "first principle": it is not universally applicable.

This raises the question whether there can be any first principles in this sense, whether mind and matter are, to some extent, subject to the same laws; or whether the resemblances which are sometimes drawn are not merely metaphors more or less expanded. Thus we speak of "weighty" objections; but will anyone maintain that ideas are subject to, or exemplify, the law of gravitation? We speak of ideas as "coherent" or "incoherent"; does anyone suppose that they stick together in the same way and from the same causes as material objects cohere?

Mr. Herbert Spencer has written a chapter[21] under the title "Society is an Organism," in which he points out many resemblances between society and an organism. But Mr. Spencer himself "distinctly asserts"[22] that the resemblances imply nothing more than that in both society and organisms there is "a mutual dependence of parts." That is to say, sociology does not herein "exemplify" some of the laws of biology, but sociology and biology both exemplify certain laws which hold good wherever there is a mutual dependence of parts.

Again, Mr. Spencer says[23] that "evolution is definable as a change from an incoherent homogeneity to a coherent heterogeneity, accompanying the dissipation of motion and integration of matter"; and, having shown how and why the homogeneous and the incoherent in the domain of physics tend to become coherent and heterogeneous, he proceeds to show that there is a similar process in the evolution of knowledge, and to say "these mental changes exemplify a law of physical transformations that are wrought by physical forces."[24] Now it doubtless is a fact that with intellectual development there goes a more accurate discrimination of differences at first unnoticed, a readier perception of resemblances at first undetected, a consequent segregation and classification of ideas, and a recognition of heterogeneity where homogeneity was at the first glance supposed to prevail. Doubtless, too, as a result of this process, there is greater coherency in our ideas. But is all this anything more than expanded metaphor? If it is something more, how can these mental changes "exemplify" a law of physics when ideas neither stick physically to each other nor gravitate to one another like the particles of the original, homogeneous, nebular mass? If mental and material phenomena can to some extent obey the same laws, and these laws are scientific inductions, there must be some resemblance between the phenomena.

That resemblance cannot be physical or spatial, because mental phenomena do not occupy space, or possess weight, or exist in three dimensions. As far as one can see, it consists in two points: both classes of phenomena (1) are objects of thought and (2) are displays of force or power. This implies that movements in space are displays of the same force as is manifested in the non-spatial movements of thought. The force which is displayed in consciousness as Will appears in space as motion. Both classes of phenomena, however, are not only displays of force, but also objects of thought. The reality of a phenomenon of either class consists in its being the manifestation of Will to or in consciousness. The reality of the two classes is co-extensive with their similarity, and is the sole foundation for any true inferences or justifiable generalisations about them.

If the law of "the instability of the homogeneous" is a first principle, and is a scientific induction based upon a real similarity between the mental and material phenomena of which it is offered as an explanation, it becomes interesting to inquire how far the similarity extends. In the case of mental evolution the essential feature of the process is that the mind gradually comes to perceive resemblances and to discriminate differences which, though they were present all the time, were not at first appreciable. In other words, the apparently homogeneous was, from the beginning, really heterogeneous. Now this fact, which is true in the sphere of mental evolution, finds its exact parallel in the evolution of the material universe, if the account given above of mechanical compounds and chemical combinations be true. What appears first in the process of evolution, and is exemplified by mechanics, is matter apparently inert. But when the particles of this apparently inert matter enter into chemical combinations with one another they reveal a fresh set of properties, quite different from those exhibited by them previous to their combination; and, when they enter into physiological relations, they display yet further additional properties. This progressive manifestation it is, and not the accompanying "dissipation of motion and integration of matter," which constitutes evolution in the material world, and finds its exact parallel in mental evolution. In both cases we have Will manifested as object of thought; and in both cases we judge most truly of that which is manifested when we judge it by its most complete manifestation. In both cases the apparent homogeneity is not the ultimate fact underlying everything, but is only the first-fruits of that which is yet to come.

FOOTNOTES:

[8] First Principles, ?46.

[9] MILL, Inductive Logic, bk. iii. ch. xii. p. 549.

[10] Ibid.

[11] First Principles, ch. iii. ?46.

[12] SETH, Chambers's Encyclopedia, s.v. "Philosophy."

[13] KIRCHOFF, Vorlesungen mathematische Physik, i. p. 1.

[14] MERTZ, History of European Thought in the Nineteenth Century, i. pp. 382, 3, note.

[15] LOTZE, Microcosmus, E. T., ii. p. 718.

[16] LOTZE, Microcosmus, E. T., ii. p. 718.

[17] MILL, Inductive Logic, bk. iii. ch. vi.

[18] MILL, Inductive Logic, bk. iii. ch. vi.

[19] CLAUDE BERNARD, quoted in L'Ann

[20] Principles of Biology, v. ?30.

[21] Principles of Sociology, vol. i. part ii.

[22] Ibid., ?260.

[23] First Principles, ?127.

[24] Ibid., ?153.

VII.

NECESSITY

We have seen that if material things can alone be treated of by science, if things which can be seen and handled are alone amenable to the methods of science, then there can be no science of mind, and no scientific laws to regulate mental phenomena. In the same way, if the field of evolution is completely filled by the redistribution of matter and motion, then there is no room left in the theory of evolution in which to accommodate the history of ideas or of morals, there is no evolution of thought or morality, no continuity between higher and lower in the intellectual development of man and the brute.

We may admit that the methods of mental and moral science, of sociology and political economy, are not identical with those employed by physics or chemistry or astronomy. But we cannot admit that the facts which, if not the proper study of mankind, are at any rate of the greatest interest to man, are not subject to or part of the process of evolution, and cannot be reduced to scientific law and order. The methods of the philosophical sciences may not be the same as those of the exact sciences; but neither are the methods of chemistry those of astronomy--just as the instruments of the astronomer are not those of the chemist. The exactness which is attained by those sciences that can apply the methods of mathematics to their subject-matter cannot be rivalled by philology or psychology. But it is not to all the material sciences that the mathematical methods can be applied: meteorology deals with matter in motion, but not yet with exactitude. The intangible and invisible, but none the less real, facts of our mental and moral experience can be measured to some extent by the statistics and averages and curves employed by the sociologist, the demographer, and political economist: the intensity of a desire may be estimated roughly and relatively by the "effective demand" for its object, the will to live by the number of suicides.

Again, we may admit that the laws of the exact or material sciences do not extend to mental science, without thereby forfeiting the right to subject mental phenomena to scientific investigation and analysis. Chemistry does not cease to be a science because chemical affinity cannot be exhibited as a case of the gravitation formula. Need psychology renounce the claim to be a science because the laws of the association of ideas cannot be deduced, say, from the laws of motion?

Of course, if science has no other object than to describe with mathematical accuracy the exact way in which material things move, if no method is scientific which does not result in such a formula, and if no generalisation, however true, is scientific which does not formulate motion in space, then, indeed, it is unscientific to talk of the evolution of mind and thought, of man and of society.

On the other hand, the movements of material things in space are facts of which we are aware, phenomena of which we are aware through our senses; in a word, they are sense-phenomena. We are aware of them as existing simultaneously and in combination, or as succeeding one upon another; and no truth, even of the most mathematical and exact of the sciences, does, or can do, more than express with mathematical exactness the precise conditions under which these sense-phenomena co-exist or follow one another, or the precise conditions without which such co-existence or sequence cannot take place. A mathematical science dealing with material things states only and always that certain sense-phenomena occur invariably and uniformly under certain conditions. The exact sciences move within the limits of the Uniformity of Nature and the law of Universal Causation; and their subject-matter consists of sense-phenomena, i.e. of things which, as known to science, are objects of perception to some mind.

But sense-phenomena are not the only mental phenomena of which we are aware: there are ideas which we do not see or handle, or smell or taste, but of which we are nevertheless distinctly conscious. Thought has its movement, ideas have their co-existence and sequences, the association of ideas has its laws. There is a uniformity of human nature as well as of external nature; there are conditions under which certain actions are always performed, and without which they would never be done. Whether the body of propositions in which these conditions are formulated be accorded or denied the name of science, matters little. But it is difficult to see what are the so great differences between these phenomena and sense-phenomena that make the latter amenable and the former insusceptible to scientific treatment. Is it that ideas are invisible? So is weight, yet the gravitation formula is scientific. Is it that thought is impalpable? So is colour, so is sound--yet there are optics and acoustics.

Be this as it may, what makes things material susceptible to scientific treatment is a quality which is not peculiar to them, but which is shared by them in common with things immaterial: it is that they are objects of which the mind is immediately aware, phenomena present to some consciousness, and that they are phenomena which appear in consciousness as co-existent and successive in certain definite uniform modes which can be detected by thought and formulated in general propositions, or laws.

If the theory of evolution comprehends all things, mind and morals as well as matter and motion; if the law of continuity connects all things together, immaterial as well as material, in a process which moves without break or interruption; it is because all things agree in the fact that they are presented (whether in sense or in idea) to the mind, and because they are presented in the continuity of consciousness.

But the object of the scientific mind is not to observe and record all the phenomena presented to it in the continuity of consciousness. On the contrary, it neglects and rejects many; but always with a purpose, viz. that of ascertaining and describing, as precisely as possible, the conditions under which a given co-existence or sequence occurs (and therefore may be expected to recur) and without which it fails to occur. In other words, science assumes that everything has a cause, and that in accordance with the uniformity of nature what has happened once will happen again in the same circumstances; that a cause will, in the absence of counteracting causes, produce its effect. Without these assumptions science cannot treat of any subject: no department of knowledge can be dealt with scientifically if these assumptions are not admitted with regard to that department. On the other hand, if by the aid of these assumptions we are enabled to reduce any set of phenomena to law and order, our success is of itself sufficient ground for regarding the assumptions as warrantable and justifiable. For science, at any rate, the only question is whether as a matter of fact they do enable us to determine under what conditions given co-existences or sequences will ensue, or what conditions such a co-existence or sequence necessarily implies.

With regard to human activity, mental and physical, it is plain matter of fact that such uniformities of sequence and co-existence not only can be but are demonstrated to prevail; and the extension of the scientific principle of cause and effect to the domain of human will and action is scientifically justified. The comparative sciences which deal with man and his works and words-- archeology, anthropology, philology--are perpetually engaged in demonstrating, with fresh proofs every day, the uniformity of human nature: in similar circumstances men have always behaved in similar ways. To satisfy the same needs, they have manufactured similar instruments at similar stages of development: flint arrow-heads from Mexico or Japan resemble those taken from British barrows; the pottery of early Greece is hard to distinguish from that of Peru; the purpose of many stone implements of unknown antiquity has been discovered by a comparison of the use to which similar tools are put by savages still existing. That man's words, as well as his works, exhibit law, order, and uniformity in their growth, as well as in their phonetic decay, is shown by the science of comparative philology. That in the face of the same problems similar analogies have been used to produce similar solutions, is revealed by comparative mythology: the imagination, which might have seemed most free to throw off the trammels of law and of monotonous uniformity, falls in similar circumstances into very similar grooves.

If the will of man is not revealed in the things which he makes, in the words which he speaks, and the thoughts which he thinks, it is difficult to know where to look for its manifestation. If, on the other hand, it is manifested in these ways, then, whether it be free or not, it is clearly uniform in its action; and the extension to it of the law of causation seems fully justified by the results.

The recognition of the universality of the law of causation must not, however, be supposed to carry with it any implication that there are no differences between, say, the organic and inorganic, or that the laws of the one are identical with or deducible from those of the other; the association of ideas may be a scientific and established fact, and yet not obey the same laws as the adhesiveness of material substances. What unites all things into a continuous, coherent, and systematic cosmos, into a scientific whole, is first the fact that, whether phenomena in sense or phenomena in idea, they are all objects of thought; and next the fact that they all exhibit the universality of causation and the uniformity of nature.

Whether this uniformity which binds man and nature into one consistent whole is a uniformity of will or a uniformity of necessity, is quite another question. It is a metaphysical and not a scientific inquiry; and the metaphysical answer, whatever it may be, is one for which science does not and need not pause. So long as nature is granted to be uniform, it matters not to science whether the uniformity is of necessity or is freely willed. In either case the sequences or co-existences described by science will continue, under the circumstances described, to happen as described.

It is, however, commonly assumed that actions which are uniform are, by their very uniformity, proved to be necessitated; and that unless what happens was bound to happen, there can be no uniformity and no science. Hence on the one hand the recognition of the freedom of the will has been denounced as fatal to all scientific conceptions of human nature; while on the other hand the uniformity of human nature and action has been denied as being inconsistent with the freedom of the will. The one side has pointed to one set of facts, which prove irresistibly that men do will the same thing under the same circumstances. The other side has pointed to the equally undeniable fact of our consciousness of freedom.

The essential feature in our consciousness of freedom is our conviction that in the present we can do or abstain from doing a contemplated action, and in the past, though we did the thing, we might have abstained from it or have done something else. Now, whether this possibility that what took place might not have taken place is a real one or only a delusion, matters not to science. If real and true, it is indeed fatal to one particular metaphysical theory, viz. that every event which ever occurred was bound to occur and could not have happened otherwise; but it leaves every truth of science, every one of those concise descriptions of what takes place under given circumstances, absolutely intact. The freedom of the will is anathematised in the name but not in the interests of science.

That becomes clear when we reflect that the laws of science are, and do not pretend to be more than, hypothetical statements. The gravitation formula does not state that bodies do as a matter of fact actually fall at the rate of sixteen feet in the first second, and so on. The statement, if made, would be untrue: a feather floats much more slowly to the ground. Still less does the formula affirm that all bodies move towards each other--and for a very good reason: many bodies are at rest. The formula makes no definite statement as to what actually does occur: it merely states what would or will happen under certain circumstances; and it is doubly or trebly hypothetical. First, it asserts conditionally that if, and only if, bodies are free to move, they will tend to move towards each other at the rate of sixteen feet in the first second, and so on. Next, even if this condition be fulfilled in a particular case, and a given body is free to move, say, towards the earth, the law of gravitation does not assert that the body will absolutely, or unconditionally, or of necessity fall sixteen feet in the first second: it only affirms that the body tends to move at that rate, and the word "tends" conveys in its meaning a second hypothesis. What is meant by saying that a body tends to fall, or tends to move in a straight line, is simply that the body will fall or move in the direction or at the rate mentioned, provided that nothing happens to prevent it. The law of gravitation then, like every other law of science, from the very terms in which it is stated, contains two hypotheses: if bodies are free to move, then they tend to move at a certain rate. Further, like every other law of science, it is based on a third hypothesis, which, as it is assumed by all scientific laws, is not expressly referred to by any. That third hypothesis is that nature is uniform: if a body is free to move it will, in the future as in the past, tend to move at a certain rate, provided that nature is uniform.

Now, throughout all this, it is obvious that science knows nothing about "necessity." Indeed, it is obvious that science, by the trebly hypothetical form of all its laws, has taken particular pains to avoid prejudging the question whether what happens was bound to happen. As we have already said, science takes care to frame its statements in such a way that they are quite independent of metaphysical theory, and will remain as true within their limits if the theory of necessity prove erroneous as they will if it turns out to be correct.

Nor can it be said, thus far, that the laws of science lead us to the theory of necessity as their logical conclusion. It may be true that if I walk over a precipice I shall fall to the bottom, in accordance with the law of gravitation. But it does not logically follow that therefore I must walk over. It may be true that a suspension bridge will fall in the same way, if the supports be removed; but it does not follow that they are therefore bound to give way. It may be true that if nature is uniform certain sequences will happen; but it does not therefore follow that nature must be uniform. In other words, the theory of necessity, if true, cannot be based on science, but must rely on some metaphysical considerations. Science does not undertake to prove even that nature is uniform, much less that it is uniform of necessity. The opposite theory, that the uniformity of nature or of human nature is due to the action of a will freely manifesting itself as uniform, may be considered superfluous from the scientific point of view. But the theory of necessity from the same point of view is equally superfluous. As long as events do happen uniformly, science has all she wants--whether their uniformity is of will or of necessity is for her quite a superfluous question. And if science were all that man wanted, these rival metaphysical theories would be of no interest to him either. But the persistency of the attempt to extract some support for the metaphysical theory of necessity out of the facts of science shows that men of science, being men, must have their metaphysics.

Are there, then, other facts of science, or assumptions essential to science, which require the metaphysical theory of necessity as their presupposition or entail it as their natural consequence? Probably the reply will be that there is one such principle: that of the Universality of the Law of Causation. The assumption that everything must have a cause may be on the part of science a pure assumption, and one which, like the Uniformity of Nature, cannot be proved by science; but it does, it may be said, assume the existence of a necessity in things.

It does, it may be replied, but whether the necessity which science assumes is the same as that maintained by the metaphysical theory in question, may be doubted. The metaphysical theory is that everything which happens happens of necessity, and could not have happened otherwise than it did. The assumptions which science makes with regard to causation are that nothing can happen unless the conditions requisite to its production are fulfilled, and that when those conditions are present the result necessarily follows. The question is whether this scientific necessity is the same as that metaphysical necessity; or, if they are not the same, whether either is a logical consequence from the other.

They are not the same: the scientific assumption is hypothetical, the metaphysical absolute. The former says that things will happen in one way, if certain conditions are fulfilled, in another if they are not; the latter that they absolutely must happen in this way, and not in that; and that it is an illusion to imagine that they can happen either this way or that. Science allows us the alternative; the metaphysical theory declares that the alternative is an impossibility or an illusion. The metaphysical theory may be right, but it is not the same thing as the scientific assumption. Neither can it be exhibited as a logical presupposition of or consequence from the scientific assumption. From a hypothetical "if" you cannot logically get an absolute "must." It may be a scientific truth that, if an electric spark is passed through two atoms of hydrogen and one of oxygen, a drop of water will be formed. But it does not follow that therefore an electric spark must be passed through them.

It is obvious that the difference between science and metaphysics in the matter of necessity is that, whereas science cautiously says, "If certain conditions are fulfilled, certain results will ensue," metaphysics boldly says, "The conditions on which the whole future depends are already absolutely fixed." Once more, this metaphysical theory may be true; but, if so, it is not from science that it derives its truth. The transition from the "if" of science to the "must" of metaphysics is illogical, though not unnatural, and is facilitated by a certain amount of obscurity, which can be thrown over it by drawing illustrations from the past. Thus, if an event has already taken place, we may infer with certainty from the fact of its occurrence that the conditions necessary to produce it were realised. And as each of those conditions must have had a cause, we can infer again that the conditions requisite to produce them were fulfilled. And so we may travel back ad infinitum along a never-ending chain of cause and effect, always moving from one fixed and necessitated event to another event equally necessitated and fixed. Thus the whole past history of the universe may be exhibited as a necessary sequence of events; and the inference may be drawn, and for the purposes of the theory of metaphysical necessity must be drawn, that because the occurrence of an event proves that the conditions required for its production were realised, therefore they and they alone were bound to be realised. Yet this is simply our old familiar non sequitur thrown into the past tense. It is true that if I walk over a precipice I shall fall, according to the law of gravitation. But I am not therefore bound to walk over. It is true that the man who fell over the cliff obeyed the law of gravity. But we cannot infer either from the law of gravitation or from the fact of his falling that he was bound to fall. We can infer that the conditions requisite to produce the fall were present, but we cannot infer from the fall that they were bound to be present. It may be quite true that they were bound to be present, but the effect which followed on them cannot be alleged either as the cause or the proof of such necessity. We must look for the reason of the necessity--if there be any necessity in the case--elsewhere. Shall we, then, say that the conditions of the fall were themselves effects of prior causes, without which they would not have happened? That again is true, but the fact that Z would not have happened had not Y preceded, is not in itself any proof that Y was bound to happen. And so we may travel back ad infinitum along the never-ending chain of cause and effect without ever finding ourselves in a position to infer from the law that everything must have a cause, that this cause was bound to operate rather than that. The occurrence of Z is no proof that Y was bound to happen, nor is the fact that Y really happened any proof that its cause X was bound to occur--and so we may work back to the beginning of the alphabet. The fact that B took place shows that A actually occurred, but not that A, rather than A>1 or A>2, was bound to occur. And if A is the beginning, what was the nature of the necessity (prior to the beginning of things) which determined in favour of A rather than A>1 or A>2?

We may indeed say, if we like--since no one can prevent us from saying things without proof or probability--that the mere fact that A happened shows that it was bound to happen. But then we might just as well have said it of Z, and saved ourselves the trouble of going through so much alphabet to get so little result. We might just as well say that as the explosion or the accident did happen as a matter of fact, it could not possibly have been prevented: Z was bound to happen under the circumstances, therefore the circumstances could not have been altered; only one result was possible under the conditions, therefore no other conditions were possible.

Or--to go back to the beginning of the alphabet once more--we may say with science that we are content with the fact that A did happen, or, since science does not profess to take us back to an absolute beginning (force and matter being eternal and without beginning), let us say we may, like science, be content with the fact that K can be shown to have happened; but whether K, rather than K>1 or K>2, was bound to occur there is nothing in science to show. If we take up this, the scientific, attitude, two consequences follow. First, there is nothing in science to require or countenance the metaphysical theory of necessity. Next, what is true of K is equally true of L or M or Z. The fact that L or M or Z occurred proves that the conditions did combine in the way necessary to produce L or M or Z, not that they were bound to combine in that way and could not have combined so as to produce L>1 or L>2, or Z>1 or Z>2 or Z>3.

Perhaps it may be said that the following is the proper way of stating the case: We have reason for believing that, as a matter of scientific necessity, if L is at work it can only produce M, and not M>1 or M>2 (the application of a light to a barrel of gunpowder can have only one result). But L was at work, therefore M alone could result. Quite true, but that does not show that the light was bound to be applied, or that the powder might not have been damp. In fine, the moment the conditions requisite for the explosion are combined, the explosion is necessary, M is the only possible result; but until then the explosion is not necessary, and the result may be M>1, or M>2, or M>3. A cause (i.e. the conditions combined) can only have one effect; but until it has that effect it is not the cause, and may never be. Pre-existent causes, which must inevitably produce predetermined effects, are figments of the metaphysical imagination. Conditions which may, and, subject to the trebly hypothetical laws of science, will combine in certain ways are scientific facts.

In fine, the Uniformity of Nature, in the sense in which Nature is assumed, both by science and by common sense, to be uniform, simply amounts to the assumption that under the same conditions the same consequences will ensue. But this uniformity neither requires nor entails necessity. The very form chosen by science for the expression of scientific laws proclaims the fact: "If bodies are free to move," "if counteracting causes be absent," "a body tends to move in the same straight line." Whatever necessity is introduced into the truths of science thus expressed is obviously imported from without, and is no part of science. We may, if we choose, read necessity into science, but there is no warrant in science for doing so. Science is absolutely without prejudice on this point. If everything that happens happens of necessity, the gravitation formula will receive no accession to its truth. If there be no necessity in the case, each and every truth of science remains valid as long as the same consequences do ensue in the same circumstances.

Since, then, science observes an armed neutrality in this dispute, and is concerned only to guard that assumption of the uniformity of nature which is vital to her existence, we must turn elsewhere for a decision of the question.

We began this chapter with an expression of our full adhesion to the view which insists upon the uniformity, not merely of nature, but also of human nature. We rejected the idea that there is no science of man, and has been no evolution of mind, as a patent absurdity, and a violent contradiction of admitted facts. Any theory of evolution and any definition of science which fails to comprehend human nature is thereby condemned as inadequate and inaccurate. For those, then, who with us accept the continuity and uniformity between nature and man there will be no difficulty in arguing from the one to the other: which of the two we shall start from will depend mainly upon circumstances, upon which is the more accessible in any particular inquiry, and which is likely to afford the best "take-off." In the present case the action of inanimate objects upon one another can be accounted for on either hypothesis, i.e. that such action is willed by some superior power or that it is necessitated by some previous action, which is necessitated by some other previous action, and so on for ever, without ever reaching any original or originating necessity. Both hypotheses will fit all the facts of all the physical sciences; both are hypotheses; and science can do and does do without either one or the other. Nor does our observation of the observed facts of nature enable us to say, with regard to any actual fact of this kind, either that it could or that it could not have happened otherwise than it did. In fine, as long as we confine ourselves to the subject-matter of the physical sciences, as long as we start in this case from nature, we cannot find anything to disturb the equal balance of the two hypotheses, which are two hypotheses and nothing more. But when we turn from nature to human nature, when we consult our own experience of our own actions, the case is notoriously different. Our experience in that case is that of two or more suggested and possible actions we are free to choose whichever we will; and our memory of past acts of choice testifies that though we actually chose one particular course, we might have abstained from it in favour of some other alternative. Here, too, as in the case of purely physical causation, the fact that a thing happened is proof conclusive that, in accordance with the law of universal causation, the conditions necessary to its occurrence were fulfilled; but it constitutes no proof or probability that the conditions were bound to be fulfilled. The fact that we chose to act in a certain way does not in the least convince us that we were bound to choose that action and that action alone. On the contrary, our memory is clear and our conviction is certain that our choice was free. In the physical and the spiritual spheres alike it is true that, when all the conditions requisite for a given effect are combined, the result must ensue. And in both spheres it is equally true that until the conditions are effectively combined no such necessity exists. In the case of our own actions we are directly and immediately conscious of the fact that it is our own will which effects this combination. For us, therefore, who hold, with Professor Huxley, that the uniformity and continuity of nature with human nature is essential to any rational and scientific view of the universe and to every comprehensive theory of evolution, it is natural to interpret physical by spiritual causation. We know from direct and personal experience of certain cases of causation that, though a particular effect necessarily ensued from a certain combination of conditions, the conditions might have been combined differently and with a different result. There is, therefore, nothing unreasonable in the inference that with regard to the events in nature the conditions which produced them might have combined differently and with different results; and that the determining factor was a will (not our own) conscious of its own freedom.

Thus far, then, the case stands thus, that in the observed facts of nature there is nothing to incline the balance in favour either of Necessity or Free-will; and that if those facts constituted the whole of our experience we should have no reason to believe the one rather than the other. But when we turn to the consideration of events of which we are the cause, we know that our contribution to the sum of conditions on which the event depends is a free-will offering which we make or decline to make as we like. That consideration would not in itself be sufficient to warrant us in inferring a similar absence of necessity in the combination of the conditions which produce natural events. If, for instance, we had reason to believe or evidence to show an absolute chasm between nature and human nature, an impossibility of their being subject to any common laws or conceptions; or if, like primitive man or the savage, we had not the accumulated observations of science to demonstrate the truth of evolution and the law of continuity--then we should have no reason or little reason, as the case might be, for interpreting nature's action and human action by one another.

The savage, as is well known, does, without any scientific authority whatever, assume straight off an entire uniformity of nature with human nature. He jumps at conclusions: he takes it for granted that everything which moves has a will of its own, like himself. But though the savage shares with the savant the impulse to believe in an essential continuity binding together man and nature, that impulse is about all they have in common. In the savage it expresses itself in an absolute identification, entirely ignoring all differences, between the two: the tree or the river has to be a conscious, rational creature, though its behaviour bears more difference from than resemblance to that of a human being. In the savant, the same impulse is trained to fertility by being constantly subjected to the guidance of observed facts: man as an animal organism is subject to the same physiological laws as other similar organisms; as an organic compound, to the same chemical changes; as a body possessing inertia, to the same physical laws. The savant's belief, however, in the continuity of nature and human nature is consistent with or rather implies points of difference between the two; e.g. man possesses a consciousness which the river does not, man is, the river is not, a conscious cause. Great though these differences be, still they are not in the eyes of science and from the point of view of evolution great enough to constitute a breach of continuity, for human actions are with the growth of science increasingly seen to be part of the uniformity of nature: the human cause only produces its effect provided that all the requisite conditions are forthcoming. Indeed, there is a danger, in some tendencies of modern thought, of ignoring the differences and of confounding continuity with identity. The distinction between the animate and the inanimate, which was hardly reached by the savage, is in danger of being overlooked by the modern materialist--an error which would be paralleled in religion by a relapse from monotheism into nature-worship. And as in the pathology of religion there is a constant tendency to substitute for religious faith a trust in the automatic efficacy of rites and ceremonies, which is a falling away into mere magic, so in metaphysic there is a tendency, in the name of science falsely invoked, to substitute for the actions of agents consciously free the operation of an automatic and magical necessity. Freedom of the will is constantly taken, or rather mistaken, both by its supporters and opponents, to mean the power of acting without a motive, and to imply that from identically the same combination of conditions one result can ensue at one time and quite a different one at another; and freedom of the will in this sense, and with this implication, is rightly rejected as inconsistent with the uniformity of nature. Freedom, however, means not the absence of motive, but the presence of more motives than one, for where there is no alternative there is no freedom, and where there is an alternative there is a choice between two things. The fact that conscious action is always action with a motive has nothing in it repugnant to the uniformity of nature, unless uniformity of nature is arbitrarily assumed to be identical with necessity. Nor has the uniformity of nature, i.e. the fact that the same action issues from the same combination of conditions, anything in it inconsistent with the freedom of the will, unless the occurrence of an event proves that it was bound to occur. The laws of science--whether physical science or mental and moral science--are hypothetical statements: if the love of gain predominates in men, then all the consequences predicted by the science of Political Economy will ensue. But this proves neither that the love of gain must nor even that it does prevail. The uniformity which marks the actions of men as often as this motive prevails is sufficient for the purposes of science, and is consistent with the freedom of the will; it does not imply that men act without a motive, nor that the same conditions produce now one effect and now another. Until the conditions which are necessary for the production of a physical event are effectively combined, physical science knows no necessity to make them combine in that particular way; if they combine in some other way, and with some other result, that combination will equally illustrate the truth of science (which says, if A then B, if $A>1$ then $B>1$), and the result will equally accord with the uniformity of nature. The same considerations apply to human nature, and if applied will be found consistent with the freedom of the will. Until the mind is made up, i.e. so long as there are alternative courses open to it, the man is free, just in the same way as in physical science, until the combination of conditions is effected, the result may or may not follow. If one alternative is adopted one set of consequences will ensue, if another, another; but whichever is adopted the results will be in accordance with the uniformity of nature, the law of cause and effect will not have been violated, the mind will not have acted without a motive, or under the influence of necessity. In fine, the universality of the law of causation lies in the fact that, however the conditions combine, each combination can only produce its peculiar effect; and whatever effect occurs can be the result only of its appropriate conditions. To say with the necessitarian that, unless at the beginning of things the course of events was unalterably fixed once and for ever, there can be no science, is to deny the universality of the laws of science, to maintain that they are true only of one particular succession of events, and would not be true of any other. In point of fact, however, the laws of science, by their hypothetical form, are adapted to cope with what is at least as striking as the uniformity of nature, that is, the diversity of nature: they apply not merely to one, but to all possible combinations of circumstances. In what way a body will move depends upon the conditions at work; but Science is not such a maimed and crippled thing that she refuses to consider its motion until she has been assured that, of the various conceivable conditions that might be brought to bear on the body, only one can, as a matter of fact, be brought to bear. On the contrary, the universality of her laws lies in the fact that they apply to all possible combinations, not merely to combination A producing B, but to A>1 producing B>1, A>2 producing B>2, and so on. The origin of all terrestrial life may be traced back, let us say, to the fortuitous combination of chemicals which constituted the first speck of protoplasm; and sundry important consequences can be shown by science to have flowed from that fortuitous concurrence. The origin of any particular species may be traced back to the accidental appearance of a sport or variety which happened to be better adapted to the environment than the parent forms were.

But, if these accidental and fortuitous occurrences had not taken place, the subsequent course of things upon earth, though there might have been no life, would still have been just as much in accordance with the uniformity (and the diversity) of nature, and equally amenable to scientific explanation. The theory that the first speck of protoplasm or the ancestral variety of a species was bound by a metaphysical necessity to occur just when and where it did, is of no use to science: if A had not happened, A>1 or A>2 or A>3 would have done, and the resulting B or B>1 or B>2 or B>3 would have been equally in accordance with the uniformity of nature and equally explicable by science.

If, then, in the physical world neither science nor the uniformity of nature requires us to believe in necessity, there is no antecedent presumption that necessity must be the law of the spiritual world: we may examine the facts of our own inner experience without prejudice. What the freedom of the will implies is that the mind has present to it more alternatives or motives than one, and that they are real alternatives and real motives, i.e. motives which may really in this particular case influence action, alternatives any one of which may be adopted in this case. The circumstances or conditions in which a man makes up his mind are, until he has made up his mind, so to speak, held in solution, and may be precipitated this way or that at his choice, or not precipitated at all, unless he chooses. The fact that in the same circumstances the same result ensues is no argument against the freedom of the will, if it be remembered that the will is itself one of the circumstances which contribute to the result, just as the mass of a body, as well as the force applied to it, helps to determine its velocity. The statement of the case then becomes this: if all the circumstances of the case be the same, and the will be the same, the consequences (i.e. the determination of the will) also will be the same. But the necessitarian position requires the statement that if all the circumstances be the same, then without any further proviso the will is determined by the circumstances; or, to put it another way, that the will does not in any way contribute to the result, which is just as though we were to say that the mass of a body had nothing to do with its velocity. But if the will does contribute to the result, i.e. to the determination of itself, it is in part self-determining.

That there must be some circumstances present, if there is to be any self-determination on the part of the will, we have already admitted; the freedom of the will implies the presence of more alternatives or motives than one--and we always have the alternative of acting or abstaining from action. But this admission only limits the powers of the will; it does not lessen its liberty. The mind can only choose between the alternatives offered to it; but as long as it has real alternatives it is free. That there must be definite circumstances if there is to be any definite determination of the will is in accordance with the fact that a cause is not some one individual thing, but a sum of conditions, every one of which is necessary to the effect, and the absence of any one of which is enough to prevent the occurrence of the result. It is a vulgar error to single out some one of the conditions (e.g. the force acting on a body) and dub it the cause, to the neglect of all the other conditions (e.g. the body's mass) which are equally necessary to the effect. It is the error committed by the necessitarian who calls the circumstances the cause, in the case of a determination of the will, and neglects the part played by the will itself.

This point of view illustrates the untenability of another objection to the freedom of the will, viz. that it implies that under the same conditions different results can ensue, or, to put it in other words, that without any change in the conditions either this or that consequence may issue. Freedom of the will is thus alleged to be inconsistent with the uniformity of nature, with the law that a cause must produce its effect. The fallacy here obviously lies in assuming that, in a modification of the will, the circumstances by themselves constitute the cause, whereas in point of fact the cause consists of the sum of the conditions, i.e., in this case, of the circumstances and the will taken in combination. Alter any one of the conditions, and the effect will be changed-- whether the condition which is changed be one of the circumstances or be the will, matters not. Conversely, if under the same circumstances a man acts one way one time and another another, the inference is not that the uniformity of nature has been violated, and that the same conditions produce different effects, but that one of the conditions was different; and as ex hypothesi the circumstances (i.e. all the conditions except the will) were in this case the same, it remains that the condition which was different in this case was the will.

Really, it is the theory of necessity which violates the uniformity of nature, for it requires us to believe that provided certain of the conditions (viz. all the circumstances except the will) are the same, then the result must be the same, no matter how much the remaining condition (the will) changes. We may, indeed, evade this conclusion by simply denying that the will is one of the conditions of its own modifications, and we may say that the wax contributes nothing to the form which it takes on when impressed by the seal. The truth is that if the will or the wax appears in the result, it must have been present and active as one of the conditions: it contributes to its own determination, and is in part self-determining.

If it be in accordance with the uniformity of nature and with our experience of what actually happens, that the circumstances should be the same and the will different on two different occasions, then the theory of necessity breaks down: if we can will and act differently under the same circumstances, we have all the freedom we want. But if--all the circumstances, save the will, being the same--the resulting modification or determination of the will is different, then the difference of result must be due to some difference in the conditions; all the conditions save one were ex hypothesi the same; the remaining condition, therefore, viz. the will, must have changed. What caused the change? Not the circumstances: one attempt to explode a barrel of gunpowder may resemble another in all the circumstances save one (the dampness of the powder), but the circumstances which remain the same (application of the spark, etc.) are not the cause of the difference in the remaining condition.

If, then, we do as a matter of fact at times under the same circumstances will different things, and if the circumstances are not the cause of the change of will, then the will changes itself, i.e. is self-determining, self-modifying. And, as we all know from experience, it determines itself at the moment of choice, not before. Until all the conditions requisite for the effect are combined, neither physical nor mental science requires us to assume that they must combine in this particular way--that the light must be applied to the powder because an explosion will take place if it is applied, that the motive of gain must be adopted because it will be gratified if it is obeyed.

Whether the conditions combine so as to produce A, or so as to produce B, the uniformity of nature is equally obeyed in either case, the law that a cause must produce its effect is equally fulfilled, and either sequence is as amenable to scientific explanation as the other. But though science and the uniformity of nature both require us to believe that when the conditions are combined the result will follow, neither requires us to assume that the combination is fixed before it is effected. And this is true equally of purely physical events and of human actions. This truth, in the case of the latter class of actions, is expressed by the statement that alternative courses of action are open to the agent, and that they are real alternatives, alternatives such that any one of them may in this particular instance be followed. From the point of view of science and of the uniformity of nature, we do not conceive that there is any difference in this respect between human actions and physical events: if science is to include both kinds of sequence and to render a rational account of them, we must assume that the principles on which conditions combine or fail to combine are the same in both cases. If physical events and human actions are both constituents in the process of evolution, there must be a continuity between them. It follows, therefore, that, in the case of physical events as well as of human actions, until the conditions are combined in such a way as to involve one determinate result to the exclusion of all others, they might combine in other ways with other results--in fine, that before the combination is effected there are always other alternatives.

At this point it becomes necessary to take into account the diversity as well as the uniformity of nature--in this case a diversity which will lead us to a higher uniformity. To the human agent alternative courses of action are open in the sense that he is conscious of their possibility and that after deliberation he adopts one or other of them. With purely physical phenomena and material things the case is different: they may be combined in this way or in that; the alternative is indeed open, so long as the combination is not effected, but it is not open to them nor is it adopted, when adopted, by them. It is adopted for them. In some cases by man. In all cases by that by which alone alternative courses of action can be contemplated and adopted--a conscious will. The course and form which man imparts to material things--to his implements or his works of art--make them so far the expression of his will; for the rest they are equally an expression of will, though of a will not his.

For those at the present day who unfeignedly accept the general principles of evolution and philosophise from them, a dualistic philosophy is impossible. They cannot hold that matter is subject to evolution and that mind is not; and the continuity of the process of evolution forbids us to suppose that there is any real discontinuity between that which appears at one stage as matter and at another as mind. There is no discontinuity if material things (i.e. the things of which we have sense-perception, but which differ from our sensations in being permanent) are on the one side the permanent expressions of Will and on the other are the transient impressions made on us in the shape of sense-phenomena.

What is true thus of the content of evolution, of that which is in process of evolution, is true also of the law of the process. We cannot suppose that it extends only to matter--that the behaviour of matter is susceptible of a rational explanation and the behaviour of mind is not. The continuity of the process excludes the possibility of a dual control: either the power which manifests itself in all things is intelligent throughout or it is not. If there is no reason in the behaviour of things, but only necessity, then those human actions and conceptions which man considers to be the result of his reason are really the result of unintelligent necessity.

It is the latter hypothesis which is expressed by the necessitarian theory. The ordinary belief of mankind--a belief which it is impossible to resist at the moment when you are making up your mind whether you will do this or not--is that you can do the thing or not, that the alternatives are real and the motives such that either of them may be acted on. The necessitarian hypothesis is that the alternatives are not real, that even before you have made up your mind there is only one alternative which you can follow--the other courses are only apparent alternatives, because you cannot choose or act on any of them; the other motives are not real motives, because by a necessity dating from the beginning of things they cannot possibly influence you on this occasion. Your action is as automatic as that of a piano which responds to the touch. The difference is that you think about the stimulus received and the piano does not. Consequently the piano makes no mistakes; you make two. You think of various possible consequences of the stimulus--which are all impossible--and you imagine that the one which you choose is the consequence of your intelligent choice, whereas it is the automatic outcome of that iron law of necessity which binds together the whole process of evolution.

It will be readily understood that a hypothesis of this kind, which is apparently in violent conflict with the plainest facts of our daily personal experience, and gives the lie to that consciousness of freedom which we all possess, would not be held in theory--it cannot be acted on in practice--unless it appeared to be the consequence of some well-established facts. It is, of course, held by its supporters to be a logical consequence from the uniformity of nature and the law of universal causation, and to be a necessary pre-supposition if we are to give any scientific account of human nature and its evolution. If, as we have argued at length in this chapter, that is not the case, if the law of universal causation only requires that a thing cannot take place unless the requisite conditions combine--and not that conditions, which did or may combine, were or are bound to combine--the question still remains, what if any value the hypothesis has on its own intrinsic merits.

In the first place it is a hypothesis which can never either be proved or disproved. The hypothesis is that our supposed consciousness of freedom is an illusion, that if we imagine we are free to choose what we will do, or that we could in the past have chosen otherwise than we did, we are deceived. The hypothesis is not based on any facts of consciousness: it is a suggestion that consciousness may be deceptive. It may: there is no means of proving or disproving the suggestion, for any reply must proceed from one consciousness to another, both of which are suspected by the maker of the suggestion to be not wholly trustworthy. We cannot ask him to concede to us, in order that we may convince him by argument, the very point which is in dispute.

In the next place, the hypothesis of necessity does formally account for all the facts which it is designed to explain: it accounts for the whole process of evolution. If everything that happens does so because it must, then the mere occurrence of any step in the process carries its own explanation with it: the mere fact that it occurred shows that it was bound to occur. If we ask, "Why was it bound to occur?" the answer is, "Because it was." Various intermediate reasons may be interpolated--because everything must have a cause, and every cause must have its effect--but if we ask, "Why must everything have a cause? why must every cause produce its effect?" the ultimate answer is always, "It must because it must." If we ask, "What proof is there that it must?" there is none. As we have already said, the hypothesis is one which does not admit either of proof or disproof.

The case is much the same with the opposite theory of freedom. Formally, the hypothesis that the whole process of evolution is throughout the expression of self-determining will is adequate to account for all the facts. But it is a hypothesis which can be neither proved nor disproved if the testimony of consciousness to our freedom may not be accepted. We cannot prove that the testimony of consciousness is true or to be trusted in this or any other matter. We take it on faith. The questions arise, therefore, Is it reasonable to take anything on faith? and if so, what? and why?

VIII.

INSUFFICIENT EVIDENCE

The theory of Design is singularly tenacious of existence, as many errors and all truths are. Science still speaks of "organs," that is of "tools" ({organa}), and of organs as performing "functions"; for the fact remains that organs are the instruments by means of which the organism acts, and that they have each their appropriate work to do, their function to perform, though science may decline to draw the inference that the instruments were designed to perform the work they do.

The Argument from Design was a comparatively simple affair as long as the organism and the environment were assumed to have been separately created: you had only to show how marvellously and perfectly they fitted one another when brought together, and it followed that they must have been designed to fit--to say they only chanced to fit was obviously absurd. But when science discovered that organism and environment were not thus independent of one another, the marvel vanished: if the environment shaped the organism, or the organism modified the environment to suit itself, no wonder that they fitted one another. It ceases to be remarkable that rivers should always flow by great cities, when we reflect that men selected sites near rivers. And chance seemed to have been established by Evolution where Design once reigned; for, if the only forms of life which can flourish in a given spot are those which are suited to the place, all we can say is that, if one form is fit to survive, it will; and if it is not, some other will. Whatever form survives will do so, not because it was designed to do so, but because it happened to be suited to its surroundings. In fine, organisms and their organs are what they are because circumstances and their past history have made them so: they have been evolved, not designed.

A little reflection, however, is enough to show that the Argument from Design is not completely excluded by evolution: things in general are what circumstances and their past history have made them; but were not those very circumstances designed to evolve what they did? Nay, are we not compelled to assume that they were so designed, if we believe in a Designer?

If, however, we ask Natural Science to discuss these questions with us, she declines the invitation on the ground that it is not her business to do so: her business is to find out in what way, not with what purpose, animal life has come to assume the various forms in which we know it; and she can do this, her business, quite well--indeed better--without discussing such questions. If it were proved that the history of animal life upon this earth had been intended from the beginning to follow the lines on which it has actually developed, not one of the problems which Natural Science has yet to solve would be brought a whit nearer solution; nor would she be any the better off, if it were proved that there was no design. She therefore very properly declines to discuss the question: there may be a Design and a Designer, or there may not; she does not know; if it is the business of science to answer the question, it must be of some other science, not of Natural Science.

So too Physical Science, when asked whether the laws of motion and matter were not designed to produce the effects which they actually do cause, replies that they may or may not, but that the law of gravitation, for instance, is equally true for her purposes, whether bodies were or were not designed to fall to the earth at the rate of sixteen feet in the first second, and so on. It may be the business of some other science to answer such questions: it is not the business of Physical Science.

And so the inquirer may go the round of the whole family of Sciences. It is an extraordinarily industrious family. It has an enormous amount of work to do: it has to feed, clothe, and generally provide for all mankind. And it can only carry on at all by a very careful division of labour: each science has her allotted task, and can only get the day's work done in the day by strictly confining herself to that task. Each science has her own questions to answer, and can only succeed in doing so by refusing to listen to any others.

The inquirer may think it strange that, in all this vast and active organisation for answering questions, no provision should be made for answering what seem to him to be some of the most important questions of all; and if he has been brought up really to believe in science, he will think it too strange to be true; he will persist with his questions, and will be eventually rewarded for his faith by discovering that there is a science which undertakes to answer them--Theology. But he will also discover that Theology is not very cordially esteemed by her sister sciences--not that they are jealous of her because she has the presumption to profess to answer questions which they acknowledge to be too high for them, but because there are grave suspicions as to her legitimacy: it is doubted whether she is a Science at all. She is, they are afraid it must be admitted, untruthful, immoral, and certainly altogether unscientific: she says what she cannot prove, and says she believes it. But they know she only pretends to believe it: they, of course, do not believe anything on insufficient evidence; what hypocrisy, then, to pretend that anyone can really believe anything except what is proved by scientific methods! They are thankful to say that they have no "faith." However, she may improve; she is certainly very backward; still, she may grow up into a common-sense science like her sisters; and then she will give up the foolish idea that she can answer questions which they cannot.

And now what truth is there in the picture thus drawn?

If there be a God, there is no other fact in the world of such awful or such blessed import to man. Religion is based on faith that there is a God. To tell the religious mind that there is no scientific proof of the existence of God is to tell it nothing new. Those were not the terms on which we took up our faith--that we should have scientific proof of everything before we did anything. On the contrary, religion begins when, and only when, a man begins "to walk humbly with his God," to know that he knows nothing except that his soul cleaves to God and humbly trusts in Him. We do not bargain so much belief, and no more, for so much proof: we give "ourselves, our souls and bodies." The gift is free. The soul shrinks from saying even that it has proof of God's existence; it only knows it hopes and longs for Him. "Faith is the assurance of things hoped for," and the strength of our assurance is as the strength of our hope. But scientific proof is not the thing hoped for: it is not what is desired when the soul is conscious of but one thing, that it thirsteth, like the hart after the water-brooks, for the living God. The humble confession of our illimitable ignorance is the foundation of our faith and will ever be its sure refuge, its inexpugnable stronghold. It is only when, being ignorant, we are tempted to deny our ignorance, that trouble begins. We drop the substance for the shadow when we believe not in God, but in some proof of God.

To the man of science all this talk about faith appears mere folly, sheer unreason, a morbid wallowing in ignorance from pure love of ignorance; and there are others who, whilst admitting that proof is not what is wanted by some minds, yet are aware, from their own sad experience, that other minds yearn for it, and can know no peace without it. And if we ask what kind of proof it is that they require, the answer is plain: it is the same kind as science insists on. Then let us go to the man of science and wait at his door: he at any rate is not ignorant, and we, if ignorant, at least are willing to learn. That he should rather look down upon us is only what might be expected in a man who by sheer force of reason has discovered the sole source of truth and built up the whole fabric of science. Certainly, when he has taken us over his palace and shown us its marvels--the balances he uses to weigh the sun, the plates with which he photographs invisible stars, the cinematographic pictures of the earth's past history, his forecasts of the future of the solar system--we are not merely willing but eager to learn how it is all done. And when we come to know him, we find that in spite of the marvels, all of his own making, by which he is surrounded, he is not puffed up, as he might have been: indeed he is, he assures us, only an ordinary man. "Scientific investigation is not, as many people seem to suppose, some kind of modern black art."[25] It is simply plain, ordinary common sense, consistently applied; and, above all, persistently declining to accept anything without sufficient evidence. In ordinary life, says the man of science, we do not swallow any statement that anybody chooses to make--we ask for some evidence; and if science waxes every day, and religion wanes, it is merely because science has made it the rule of her being never to believe anything without sufficient evidence, and religion has not.

Naturally, then, we wish to know what is "sufficient evidence" in the eyes of science, since everything, we are told, depends on that. The reply is brief: whatever is based on the Uniformity of Nature has sufficient evidence. If we are inclined to be puzzled by the "Uniformity of Nature," we are soon reassured; it is literally the most ordinary thing in the world, there is no difficulty about it. Man is born into a world in which changes are unceasingly taking place. Some things change even as the clouds shift--every second, and in a way patent to all beholders. Others change imperceptibly and with great slowness, as e.g. the level of the dry land or the shape of the coast-line. But all things change, {panta rei}. Nothing abideth long in one stay. It is these changes which bring all things good to man, and also all things ill. If, then, man is to survive, he must learn to evade the latter changes, which threaten to crush him, and he must be there in time to profit by the former. Such was the problem presented to primitive man, and such it still is for every one of us to-day: the successful man is the one who is beforehand with the world, and, if he is beforehand it is because he has learned to read the signs of the times and the seasons. In a word, he has learned to recognise that changes are not always mere chances, that some changes are uniformly preceded by certain others, and may consequently be foreseen. In the beginning the changes that man can forecast are few indeed: his prevision is no greater than the brute's. The child does not foresee that fire will burn; he learns by experience. And whatever man can forecast, he has learned it all by experience. It is a slow way of learning, it has taken man thousands upon thousands of years to learn what he knows now; still he has learned to know the causes of countless things, to control the causes and to anticipate the effects of many. But more important, more valuable than all his experience and all his knowledge of what produces what, of what uniformly precedes or follows what, is the final and comprehensive truth which at last he reaches, that nothing happens arbitrarily, that everything in nature is uniform. That, the Uniformity of Nature, is the great truth in which all others are summed up: to its establishment have gone the labours of all past generations of mankind, to its support the whole experience of the race contributes. It is the truth of truths, the test of truth: whatsoever is established on it shall not be shaken, whatever contravenes it shall not endure.

The Uniformity of Nature is the base not only of all science, but of every act of reason in the most commonplace affairs of ordinary life; and, though you may not know it, you assume it every moment. Why are you sure that the sun will rise to-morrow? Because Nature is uniform. Why do you know that fire will burn? Because Nature is uniform. Why that all men are mortal? Why that a cause will always produce its effect? Because of the Uniformity of Nature. For each and all of these beliefs the evidence is sufficient; it is the Uniformity of Nature. How different, says the man of science, is the procedure of science, that is of common sense, from the unscientific methods of theology! Why do we believe that the earth will bring forth her kindly fruits in due season? Because it is God's will? That is a hypothesis; it may be true or it may not; it cannot be proved or disproved; there is no evidence against it, but there is no evidence for it. Very different is the answer of science and common sense: it is that the earth will produce crops in accordance with certain natural causes, mechanical and chemical. That also is a hypothesis which may or may not be true. Yes, but it is one for which there is some evidence--the Uniformity of Nature. In the same way, if anyone were to say that the result of the next general election depended not on the electors but on the planets, we should decline to believe him, because there is no evidence to show that the planets have anything to do with it, and there is good evidence for believing that the votes of the electors have. In fine, the teaching of science is: demand sufficient evidence for everything, and always remember that by sufficient evidence is meant the Uniformity of Nature.

This sounds so simple and so convincing that we are tempted to try it. But first let us make sure that we have learned our lesson properly. In the course of long ages mankind has slowly accumulated enough experience to warrant the confident belief that Nature is uniform. Now, primitive man was of course a savage, and knew nothing of the Uniformity of Nature; he therefore could not have had sufficient evidence for believing anything in his experience. But it is on the accumulation of such experiences--every one of which we must reject because they were not based on the Uniformity of Nature--that our belief in the Uniformity of Nature is supposed to rest. In other words, it is based on them and they were based on nothing. This result of acting strictly up to the principle of not suffering anything to pass without sufficient evidence seems somewhat discouraging, until the man of science comes to our rescue and reminds us that just as we, without knowing it, have acted all our lives on the tacit assumption that Nature is uniform, so did primitive man; and that consequently there really was sufficient evidence and scientific proof for the savage's experiences, though of course he could not have framed it in words; and so, the bases of the Uniformity of Nature are really quite sound. But even now we are not altogether out of our difficulties, for granted that the savage, like ourselves, tacitly assumed Nature to be uniform, was there sufficient evidence for the assumption? and if so, what was the evidence? It could not be the Uniformity of Nature, because that is just the question; and, if it was anything else, it was not sufficient evidence.

It really seems rather difficult to get sufficient evidence for the axiom, viz. the Uniformity of Nature, on which the whole of science is built. And yet we must have sufficient evidence for it, or else we shall have to conclude that Science has no more logical foundation than Religion.

But once more the man of science comes to our assistance and explains that in the beginning, before the Uniformity of Nature is proved, it is only probable that what has once happened will happen again in similar circumstances, and at first perhaps not very probable; but when wider and wider experience still shows that what has once happened does actually happen again under the same circumstances, the Uniformity of Nature becomes more and more probable, until at last, if not actually proved, it is still the most probable hypothesis that we possess or can possess: "our highest and surest generalisations remain on the level of justifiable expectations; that is, very high probabilities."[26]

Now, with all respect to logicians like John Stuart Mill, and men of science like Huxley, we must point out that this begs the whole question. If we assume that Nature is uniform, then it is probable that what has often happened will happen again. But if we do not assume that Nature is uniform, then the repeated occurrence of a thing does not make it in the least probable that it will occur again. To assume without proof that Nature is uniform is to ask us to accept a statement without evidence, which, if we have learnt the lesson of science, we can hardly do. On the other hand, if we begin with the admission that Nature may or may not be uniform, but that, to begin with, we no more know whether it will actually prove to be uniform than we know whether a penny, when we are about to toss it, will fall head or tail; then, according to the mathematical theory of probability, it matters not how many times you toss the penny, the chances next throw are exactly the same as they were at the first throw--it matters not how many times Nature has proved uniform in the past, she is no more likely to prove uniform to-morrow than she was on the first of days. If it is really an open question at the beginning whether Nature is or is not uniform, it remains an open question to the end. The man of science need not admit that it is an open question, if he does not want to do so; but if he does admit it, then let him stick to it throughout; and let him reflect that if he begins by admitting it and ends by denying it, he has but gradually retracted his own free admission, and unconsciously been betrayed into denying what he began by admitting to be true.

The fact of the matter is that the axioms of science--the Uniformity of Nature and the Law of Universal Causation--not only are not proved by what experience we have had of them, but "cannot be proved by any amount of experience."[27] Not only can they not be proved by any amount of experience, they are incapable of being demonstrated at all: "they are neither self-evident nor are they, strictly speaking, demonstrable."[28] If, then, they are not and cannot be proved either by experience or in any other way, on what does the man of science ground his belief in them? On Faith. "The ground of every one of our actions, and the validity of all our reasonings, rest upon the great act of faith, which leads us to take the experience of the past as a safe guide in our dealings with the present and the future."[29]

FOOTNOTES:

[25] HUXLEY, Darwiniana, p. 361.

[26] HUXLEY, Science and Christian Tradition, p. 205.

[27] HUXLEY, Evolution and Ethics, p. 121.

[28] HUXLEY, Method and Results, p. 61.

[29] HUXLEY, Science and Christian Tradition, p. 243.

IX.

CONSEQUENCES

In the last chapter, impressed by the doctrine that there is no "source of truth save that which is reached by the patient application of scientific methods,"[30] we patiently applied those methods to the foundation of science itself; and we were rewarded by the discovery that scientific, like religious, truth has its source in Faith. But the end of our difficulties is not yet.

A man may put his faith in science, if he will, "but let him not delude himself with the notion that his faith is evidence of the objective reality of that in which he trusts."[31] About that we feel no difficulty: faith begins not merely with ignorance, but with the frank confession that we know we are ignorant, but we wish to believe, in spite of the absence of evidence. There is no evidence to show that Nature is uniform or science true, but we do not mind that: we are quite determined to believe, evidence or no evidence. That is easy enough for us, who are not scientific; but "scientific men get an awkward habit--no, I won't call it that, for it is a valuable habit--of believing nothing unless there is evidence for it; and they have a way of looking upon belief which is not based upon evidence, not only as illogical, but as immoral."[32] This is, if not awkward, at least puzzling, since science is based on a belief in the Uniformity of Nature, for which there is no evidence.

"It is, we are told, the special peculiarity of the devil that he was a liar from the beginning. If we set out in life with pretending to know that which we do not know; with professing to accept for proof evidence which we are well aware is inadequate; with wilfully shutting our eyes and our ears to facts which militate against this or that comfortable hypothesis; we are assuredly doing our best to deserve the same character."[33] That also is puzzling. Science sets out in life with assuming, by a "great act of faith," that Nature is uniform. She is well aware that the evidence for this assumption is inadequate, that no amount of experience could prove it; but, if she is to start at all, she must make the assumption, so she proceeds to act as though it were proved, as though she knew what she does not know. These are facts; and we take it for granted that no one will wilfully shut his eyes and his ears to them, even if he has some comfortable hypothesis against which they seem to militate.

Again, belief in science is based not on any ground of reason, but upon "the great act of faith" which leads the man of science to assent to it. It is therefore again puzzling to learn that "assent without rational ground for belief is to the man of science merely an immoral pretence," and that "scepticism is the highest of duties; blind faith the one unpardonable sin."[34]

But the reader has probably already correctly divined the solution of these puzzles: the passages quoted above are not intended to apply to science. The blind faith which is illogical, immoral, a pretence and a lie, is, of course, not faith in science, but some other kind, which may therefore be dismissed; and we may start once again with the happy feeling that there is one kind of faith at least which is logical, moral, and real and true.

It is, then, quite honest and logical to have faith sometimes; and, without evidence, to believe some things, e.g. the Uniformity of Nature. Here, however, some readers may interpose with the objection that the man of science has not proved that his faith is logical and moral, and real and true--he has simply assumed it. Quite true; but that is his faith and we must respect it, as we respect any man who holds fast to what he honestly believes to be the real truth. We do not imagine he could believe it if he thought it a pretence or a lie. And we do not call upon him to prove it before we believe him--still less to prove it before he believes it himself.

It is, therefore, we repeat, quite reasonable to believe in the Uniformity of Nature without evidence. The reluctance that is genuinely felt by many minds to take up this position is probably due to a feeling that if we may believe in one thing without evidence, then anyone may believe in anything he likes. And it would not be quite fair to make the rejoinder, What does that matter to you, as long as you are free to believe what you think right? The tendency to dogmatise, and to be intolerant of opinions not our own, is, indeed, strong enough in all of us to make us stand somewhat in dismay of a line of argument which seems to indicate not merely that other people have a right to differ from our opinions, but may quite conceivably be right in so differing. Still, this tendency does not wholly account for our reluctance. That reluctance has, in part at least, a nobler origin than narrow-mindedness and the ignorance which knows not that it is ignorance. It does matter to us what our fellow-men believe. Still more does it matter how and why they choose their beliefs.

The reluctance to admit that it is permissible to believe without evidence even in a truth so undisputed as the Uniformity of Nature, is also in part due to yet another cause. It is felt that to admit belief without regard to evidence is to invite intellectual anarchy, and to leave mankind the helpless prey of ignorance, error, and superstition. Hence, in many candid souls, a lamentable feeling of distraction and hopelessness: to abandon their old faith, even if it has no evidence, is almost more than they can bear; to retain it, knowing that it has no evidence, is to open the floodgates of a saturnalia of unreason by which the foundations of civilisation would be swept away. Hence, too, the zeal with which other minds call for the destruction of every belief, but especially religious belief, not based on evidence, and with which they denounce faith as the one unpardonable sin.

But the error into which both classes of mind fall is a simple one. It consists in imagining that if we take one thing on faith, because there is no evidence, therefore we may believe anything, even if the evidence is conclusive against it--that if we once accept faith, we must for ever abjure reason. The error has been clearly exposed by Professor Huxley, who, after pointing out that reason--ratiocination--is based on faith, says, "But it is surely plain that faith is not necessarily entitled to dispense with ratiocination because ratiocination cannot dispense with faith as a starting-point; and that because we are often obliged, by the pressure of events, to act on very bad evidence, it does not follow that it is proper to act on such evidence when the pressure is absent."[35]

It seems, then, a piece of alarmist exaggeration to say that if we admit one thing, e.g. the Uniformity of Nature, without evidence, we forfeit the right ever again to ask for evidence for any other statement: on the contrary, whenever evidence can be got, we must get it and abide by it. But this only shows that no disastrous consequences will necessarily ensue, if we frankly admit what in any case is the fact, viz. that there is no evidence for the postulate on which all science is built. You will not have committed high treason against the best interests of mankind by acting, in this case, on the principle that a man may sometimes believe a thing on evidence which, he is well aware, is insufficient, or on no evidence at all. On the other hand, in another case, to act on the principle might be, if not high treason, at least mischievous.

It seems, then, first, that there are some things which a man may believe without evidence, and some which he may not; and, next, that he may not believe things the consequences of which would be disastrous or mischievous. But now what of the things not mischievous or disastrous? On what principle are we to choose amongst them? Let us once more follow our guide, the man of science, and ask him on what principle he elected to believe that Nature was uniform, rather than that she was not. I imagine it was once more on the ground of the consequences: grant that Nature is uniform, and then all the marvellous discoveries, the revelations of the past and prophecies of the future, which science has made, become things that we can reasonably believe in. Refuse to believe, withhold your faith, and then you have no reason to believe anything whatever, thought and action alike are paralysed. It is between these consequences that we have to choose. Our choice is an act of will; and it is on our will that our beliefs and our actions depend.

In science, then, we are offered the alternatives: either believe without evidence that Nature is uniform, or renounce all that science has to give. We want to be scientific, so we choose the former. We believe (in science) because we want to believe, not because we have any evidence. To say that we may yield to the impulse to have faith, without being unscientific, is to understate the case: we cannot be scientific without faith.

In logic, whether inductive or deductive, the case is the same. We must either believe without evidence in the axioms on which reason is based, or forego reason altogether. We want to be reasonable, so we choose to accept the axioms. But our choice is not the least evidence or proof that they are true. We believe they are true, because we wish to believe that they are true. There is no reason except there first be faith.

With morality the case is not otherwise. We believe in the principles of morality, not because we can prove them, or bring evidence to show that a man ought to do what is right, but because we wish to believe, and because we have faith in the right. There is no morality except first there be faith.

We are nothing, know nothing, can do nothing without faith. And it is not in the dead past, which is what we mean by "evidence," but in the living future that faith has its well-springs. It is because we wish to do right henceforth that we put our faith in right-doing. It is not the ghosts of our misdeeds, rising from the charnel-house of the past in evidence against us, that give us good hope of the future--it is faith, not built on evidence, on a past that cannot be altered, but on hope, on the future, on what shall be as we will it.

The future is uncertain. But that is no reason why you should be. There is no evidence that we shall succeed, that logic can be trusted, or that science is true. But fortunately it is possible to be certain without evidence. In commenting on the text "Faith is the assurance of things hoped for, the proving of things unseen," Professor Huxley says, "I fancy we shall not be far from the mark if we take the writer to have had in his mind the profound psychological truth, that men constantly feel certain about things for which they strongly hope, but have no evidence, in the legal or logical sense of the word; and he calls this feeling 'faith.'"[36] It is a profound psychological truth, and by the aid of the theory of evolution we may understand why it is so deep-seated in the mental and moral constitution of man. Primitive man can have had no extensive "evidence" of any kind to go upon in regulating the conduct of his daily life; and in all probability exercised but little power of criticism in judging the value of what evidence he had. At the same time, if he was to survive at all in the struggle for existence, he had to act and to act promptly. Fortunately for him it was possible to feel certain about things for which there was no evidence, i.e. to have faith. And he survived in consequence--in virtue of the law of the survival of the faithful, a law whose operation is possibly not confined to this world.

On the theory of evolution, again, man's wants must have aided him in the struggle for existence; and no evolutionist will doubt that the desire to be rational and to do that which is right has assisted man in his upward struggle. The victory has remained with those who have been contented to feel certain about things for which they had no evidence, and to act on faith. It is those who hesitate to do right until sacrifice of self is proved to be reasonable, who lose their chance, and consequently have been and are being, though slowly, weeded out. Those who have yielded to their inner impulse to believe, without evidence, have evidently been the better fitted to their environment, and the more in harmony with the ruling principle of the cosmos and its evolution.

Thus far in this chapter there has been no explicit mention of religious faith. We began with the fact that faith is indispensable to science as its starting-point. We do not wish to end with the suggestion that scientific faith can or ought to be stretched so as to make religious faith its logical or necessary consequence. On the contrary, the man who by a great act of faith accepts the

Uniformity of Nature without evidence, and then resolves never to accept another statement without evidence, is quite safe: no one can make him believe in religion as long as he holds to his resolve--or in morality either. There is no evidence--and therefore he cannot believe--that a man ought to do what is right. If he does ever depart from his resolve as regards morality, it will be because in his heart--with its reasons which his reason knows not of--he wants to do right, not because there is any evidence.

In most men the impulse to believe expends but does not exhaust itself in reason and morality. There is also the religious belief that all that happens to us is due to a Will not our own, in which we can trust and to which we can give our lives. For this belief there is no more evidence than there is for science: if a man will receive it, he must believe in it as he believes in science, that is, without evidence. If a man will receive it, he may, on the same condition as he believes in morality or science, viz. that he wants it. Any other condition is of his own making and is an act of his own will: if he says that he fain would believe, but cannot without evidence, that is a condition of his own making, imposed upon him by his own will--what science and morality both require cannot be immoral or unscientific, and they each require belief without evidence in order that they may exist at all. What logic postulates can hardly be illogical. It can be no necessary law of reason to check the impulse which gives to reason its initial impetus. We believe that science is true for no other reason than that we wish it to be true; and for every man, with regard to religion, the question is, does he wish it to be true? if it lay with him to decide, would he have it true? if he would, then it does lie with him to decide: let him be assured it is true. If he would not, let him ask his own heart, Why? Why does he wish there were no God?

FOOTNOTES:

[30] HUXLEY, Science and Hebrew Tradition, p. 233.

[31] HUXLEY, Science and Christian Tradition, p. 245.

[32] HUXLEY, Science and Hebrew Tradition, p. 65.

[33] HUXLEY, Science and Christian Tradition, p. 54.

[34] HUXLEY, Method and Results, p. 40.

[35] Science and Christian Tradition, p. 243.

[36] Science and Christian Tradition, p. 244.

X.

THE CHESS-BOARD

We began, at the beginning of this book, by accepting evolution as a fact, and by asking the question: Granted that it is a fact, what follows? What does it mean for me? What light does it throw on the meaning of life?

The answers that we may give to these questions together constitute a philosophy of evolution, which is carefully to be distinguished from evolution as a scientific theory. As a scientific theory evolution is an account, as exact as science can make it, of what actually did happen in the past, of the precise process by which things have come to be what they are. When this knowledge has been gained, we may ask the question, What value has this knowledge for the practical purposes of life? And the answer will be a contribution to philosophy, but it will not be one of the things described by science as having happened in the past, will not be part of the knowledge from which it is itself inferred, nor, if it is a false inference, will it have any right to masquerade as science and say that we must accept it as true or else deny the truth of science. Indeed, we found that two answers to the question, two philosophies of evolution, the Optimistic and the Pessimistic, have been formulated, which being contradictory cannot both be true, though both may be false.

The Optimistic theory, that evolution is progress, only established its conclusion, that the process of evolution is necessarily from good to better, by means of arguments which denied the distinction between good and bad, and implied that our moral convictions were illusions.

The Pessimistic theory, on the other hand, assumed the reality of our moral ideals, but was forced by its adoption of the theory of Necessity to conclude that it is an illusion to imagine those ideals can be finally realised.

Both philosophies in theory profess to make no assumptions, to take nothing on faith, and to base themselves on nothing but what we actually know to be facts. In practice each of them does unconsciously base itself on faith and does tacitly make certain assumptions. But as the assumptions made are not precisely the same in both cases, they reach two very different conclusions--Optimism and Pessimism. Again, if each philosophy treats as illusions certain facts--the freedom of the will and the reality of moral distinctions--which the common sense and common consciousness of mankind hold to be real, it is because each philosophy arbitrarily rejects certain of the assumptions which common sense makes, certain articles of the common faith of mankind. Consequently, when we find that each philosophy is inconsistent with itself, and ends by implying that what it assumed to be real is in fact an illusion, we are led to suspect that its assumptions may not have been adequate or well-considered, its faith not great enough to remove mountains or explain the world.

The conception of a "positive" philosophy--that is, a philosophy which confines itself to positive facts, and which is "agnostic" in the sense that it does not profess to know what it knows it does not know--is borrowed from science. It is an attempt to carry the methods of science into the domain of philosophy, to substitute science for philosophy. The attempt is made under the impression that science does not profess to know what it knows it does not know, i.e. makes no assumptions and takes nothing on faith. That impression, however, is, as we have argued in the last chapter but one, a false impression: the Uniformity of Nature is a pure--and rational--assumption. If, therefore, a philosophy confined itself strictly within the bounds of science, it would not be strictly positive or agnostic: it would still make some assumptions, even if only those made by science, and would still, even if it confined itself to the positive facts of science, be taking something on faith. A sound philosophy is one, not that makes no assumptions, but which seeks to find out what assumptions are made by any department of knowledge or practice--science, art, evolution, morality, religion--and how far those assumptions will carry us. The bane of philosophy is not making assumptions--all thought does--but is thinking you have made none.

Common sense assumes that the testimony of consciousness, so far as it can be verified by consciousness, can be trusted as evidence of the reality of that which is presented to it. Positive or agnostic philosophies, whether of the optimistic or the pessimistic type, on the principle of making no assumptions, reject this one, either on the ground that the Real is Unknowable (which is itself an assumption as incapable of proof or disproof as the assumption that the Real is Knowable) or on the ground that we only know our states of consciousness, and cannot know whether there is or is not any reality beyond them (which again is simply an assumption that consciousness as evidence of a reality beyond itself is not to be trusted).

Now, granted that common sense makes an assumption here, as it assuredly does, it is one such as can only be rejected by making a counter-assumption: to refuse to trust consciousness as evidence of a reality beyond itself is to make the assumption that it is not trustworthy--which may or may not be true, but is just as much an assumption as the supposition of its trustworthiness is. The positive and agnostic philosophies, therefore, do not succeed in avoiding assumptions in this matter: they only tacitly add another to that which they have already unconsciously made by assuming that Nature is uniform.

If, now, they adhered to these assumptions, we might proceed to ask what conclusions they deduced from them. We should not, indeed, expect their conclusions to be the same as those reached by persons starting from the opposite hypothesis, viz. that consciousness is trustworthy. And we should not agree that they were superior to those reached by the common sense and drawn from the common faith of mankind. We should only admit that they were different, because drawn from different premises. The argument that the teaching of a philosophy which makes no assumptions must be superior to one that does, is an argument which, whatever its value, we should have to set aside in this case, on the ground that the agnostic philosophies are not so ignorant as they modestly profess to be: they do know something--they know that Nature is uniform, and that consciousness as evidence of reality is not to be trusted--or they assume they know.

But the positive philosophies do not adhere to their assumptions. Few philosophers do. The optimistic evolutionist takes back his remark about the untrustworthiness of consciousness, so far as material things are concerned: matter and motion at any rate are real, and consciousness is good evidence, as good as can be got, of their reality. The pessimistic evolutionist also repents him, as far as our moral convictions are concerned: they are fundamentally real; our consciousness of the moral ideal is our best evidence for it.

On the other hand, both the optimistic and the pessimistic evolutionist adhere with perfect consistency to their rejection of the evidence given by consciousness to the freedom of the will. But here, too, the assumption of common sense cannot be rejected without a counter-assumption: if it is a pure assumption to say that things could have happened otherwise than they did, it is equally mere assumption to say they could not.

Finally, there is one other assumption made by the common faith of mankind and rejected by positive philosophies. It is that the world, i.e. everything of which man's consciousness is aware and to the reality of which his consciousness is evidence, is the expression of self-determining will, human and superhuman, manifesting itself directly to his consciousness. This assumption, too, has its counter-assumption--that there is no self-determining will, human or superhuman--and to reject the one assumption is to accept the other. To say that you do not know whether a man's word may be trusted or not is literally agnosticism, and may be the only rational attitude to assume, e.g. if the man is an absolute stranger, as most witnesses in court are to the judge who tries the case. But on the ground of your ignorance to refuse to pay any attention to his evidence when given is to abandon your agnosticism--if a judge directs the jury to disregard the evidence of the witness, the presumption is that he assumes it to be false. So, too, if we disregard the evidence of consciousness on this or any other point, we do not thereby succeed in avoiding assumptions, we only assume that consciousness is not trustworthy.[37]

The idea that in philosophy it is possible permanently to maintain an agnostic attitude with regard to the trustworthiness of consciousness is the outcome of a conscientious attempt to apply scientific methods to the solution of philosophic problems. Science does not find it necessary to assume either that there is or that there is not a God: on either assumption it is certain that bodies tend towards each other at the rates specified in the gravitation-formula. Philosophy must be made scientific. Therefore philosophy must carefully avoid making either assumption. Why, the very reason why science has progressed and philosophy never moves is that science builds only on demonstrated fact, philosophy only on undemonstrable assumptions. Proof, and therefore truth, is impossible if you start from assumptions which never can be proved to be either true or untrue.

The truth is that it is possible to maintain the agnostic attitude, and to avoid making assumptions, just so long as we do not need to form an opinion or take action on the matter which the assumption affects.

If my interests, practical or speculative, are not affected by a certain trial now proceeding in the law courts, I can avoid making any assumption as to the trustworthiness or untrustworthiness of a witness's evidence. I do not know whether he is trustworthy or not, and I can refuse to make any assumption whatever on the subject--there is no reason why I should. But the moment circumstances call on me to form an opinion, I find myself beginning to make one assumption or the other, or perhaps at first one and then the other, though I am just as ignorant whether he really is trustworthy or not as I was when I refused to make any assumptions; I know no more about his previous career or his antecedent credibility than I did before he entered the box.

So, too, the truth is not that science makes no assumptions, but that she makes no assumptions except those which are necessary for her purposes. The man of science assumes--and it is pure assumption--that he can trust the evidence of his consciousness as to the reality of the chemicals he experiments on, the plants he classifies, or the stars he observes. He assumes that they are real. He also assumes without proof that what has produced a certain effect once will produce it again in the same circumstances, that if a thing has occurred the conditions essential to its occurrence must also have occurred--in fine, that Nature is uniform. But so long as he is engaged exclusively in scientific work, in finding out what actually does happen or has happened in Nature, he need make no assumptions as to whether a certain witness is trustworthy or not, or whether there is a God or not: he can maintain a perfectly agnostic attitude on both questions. He can say, if he chooses, "God or no God, two and two make four"; or, to put it more precisely, whether the evidence which consciousness gives in spiritual experience to the reality of a God can or cannot be trusted, I do trust the evidence which consciousness gives in sense-experience to the reality of material things; whether the assumption that every event is the expression of self-determining will is true or not, at any rate I believe in the assumption that Nature is uniform.

And if man had nothing to do but investigate the actual course of Nature, and had nothing else to form an opinion about except whether this phenomenon is followed by that, it would be possible permanently to avoid making any assumptions save those required by science. But man has (let us suppose) to know, not only what does happen, but what ought to happen, and to decide what shall happen. The ordinary man, in making those forecasts of the future which he must make for the ordinary business of daily life, assumes, quite unconsciously, that Nature is uniform and that material things are real. In deciding what he ought to do and what he will do, he assumes, without knowing that he is making any assumptions at all, that his moral ideals are real, that his will is free to choose this course or that, and that the God with whom he communes in his heart is real.

Let us now take the question raised by agnosticism as to these assumptions, which constitute a large part of the common faith of mankind. The question is not whether these assumptions are right: the agnostic declines to discuss that question; he does not know whether they are right or wrong, he has no means of deciding, they are too high for him. The question is whether the agnostic himself succeeds in making, as well as endeavouring to make, no assumptions on these points. We have already argued he fails: he succeeds, not in making no assumptions, but only in making the counter-assumptions to those assumed by common sense.

This, let us hasten to add, does not at all amount to saying that his counter-assumptions are wrong: it only amounts to saying that he cannot form a resolution to steal or not to steal, to lie or not to lie, without (consciously or unconsciously) making some assumption as to the reality of the moral ideal.

But there is little need of argument to show that agnostic philosophy fails to avoid making assumptions, i.e. fails in practice to be agnostic. Professor Huxley admitted that the Uniformity of Nature was an assumption; he assumed that our moral ideals were real; he took it for granted that the will was not free. We need only point out that the attempt to carry on philosophy and explain the universe on purely scientific principles breaks down: science makes no assumption about the reality of our moral and aesthetic ideals; philosophy, even an agnostic philosophy, finds it necessary to assume the reality of both. Even if philosophy could be made scientific, it would not get rid of unprovable assumptions: it would still be based upon those made by science. And the excellence of philosophy, or of any explanation of the universe, consists, not in agnosticism, not in making no assumptions, but in making the right ones.

Science, as we have said, makes no assumptions save those which are necessary for her purpose, which is to ascertain and describe what actually takes place in Nature. Conversely, it is vain to imagine that from those assumptions anything can be deduced except conclusions of the kind which they are framed to cover, viz. conclusions as to what actually does take place. To say, therefore, that all knowledge--philosophy and religion--must become scientific before it can be regarded as trustworthy is simply to say that nothing can be regarded as true, except what is deduced from the assumptions of science: conclusions drawn from any other assumptions have no scientific truth. The assumptions of science are constructed only to lead to conclusions as to what is: we can therefore have no scientific, i.e. no real, knowledge of what ought to be. With the assumptions she makes, Science can only describe the way in which things happen; why they should so happen it is therefore impossible to know. The idea that all things are the expression of self-determining will is not one of the assumptions of science; no conclusions from it, therefore, can be considered valid.

Without staying to consider why the unproved and unprovable assumptions of science are so superior to all others as to be set up as the sole source of truth, the only fount of genuine knowledge, let us consider what sort of a picture of the universe they give us. Perhaps a simile will best help us.

Let us imagine a game of chess in course of being played by invisible players in presence of a scientific philosopher who knows nothing about the game--or who assumes that he knows nothing--except what his senses tell him.

What he sees will be simply material chess-men moving in space. He may either consider them to be merely sense-phenomena, merely affections or modifications of his sense of sight and touch, or he may consider them to be real, material things. In either case he makes an assumption. The latter assumption leaves it quite an open question whether the reality is something insentient or is the expression of conscious will. The former precludes the question, i.e. assumes that there is neither conscious will nor insentient matter behind them.

But in neither assumption is there anything to prevent the philosopher in question from studying the movements of the chess-men and the way in which at every move or moment they are redistributed. At first their movements would probably be rather bewildering; but in course of time he would note, we may assume, that Black never moved unless White had previously moved, and that any movement of White was followed by one on the part of Black. He might therefore be tempted to lay it down as a rule that Black never moved unless White moved first--that an effect never occurred without a cause; and that a movement of White was always followed by a move on the part of Black--that a cause was always followed by its effect. But if he yielded to this temptation he would be making an assumption, for--inasmuch as he professes to know nothing to begin with--he does not know that the pieces always will move in this way; he only knows (assuming that memory is not a mere delusion, as it may be, for anything he knows) that they have moved thus, not that they always will move thus. He may, however, assume that they will continue to move in that way. But with every fresh assumption he becomes less and less of an agnostic. He may, indeed, if he likes, further assume, not only that the pieces will move in this way, but that they must. This assumption does not, indeed, seem necessary; for if we know (or assume that we know) that they will follow this course, it seems superfluous to say that they must.

It seems well, therefore, to try to see on what principle we are to make our assumptions. It is an ancient rule, and one followed by science, to make as few as possible--that is to say, the fewest that will suffice for the purpose in hand. If, therefore, the purpose of our study of the chess-board is merely to find out how and according to what rules the pieces actually do move, have moved, and will move, it seems sufficient to assume that they will move as they have done, not that they must. If, on the other hand, we want to know why they move in the way that we assume them to move, then the assumption that they do so because they must is certainly in form legitimate, though it may or may not be the right one in fact.

Some people refuse to discuss such questions as "Why this universe?" "What is the reason of this unintelligible world?" on the ground that they cannot be answered except by making assumptions which cannot be proved.

But is that really a good reason for refusing? If it is, then none of the questions which science exists to answer can be discussed, for they also can only be answered by assuming, without proof or possibility of proof, that

Nature is uniform, that the chess-men will continue to move as they have done.

Be this as it may, our philosopher, if he assumes that the course of Nature is not only uniform, but necessary, is making an assumption which is not required for the purposes of science, though it may be for his philosophy. It is, as we have said, quite legitimate for him to make the assumption for philosophical purposes, and to adhere to its logical consequences. But in the interests of clearness of thought it should be recognised that those consequences flow from it, and not from any of the assumptions necessary for the purposes of science. He will be able to show on this assumption that there is nothing in the history of the universe, or in the facts of science, to countenance the idea that the universe is the expression of self-determining will. We only wish to point out that this conclusion, even if true, is not an inference from the facts of science, but from the initial assumption that nothing which takes place in Nature is the result of free will.

To say, "Science does not find it necessary to assume the existence of self-determining will, neither therefore will I assume it," is true, but is only half the truth. Science does not find it necessary to assume the non-existence of self-determining will. But the philosopher who explains the facts of Nature on the hypothesis that they happen of necessity, does assume that self-determining will is non-existent. It is therefore quite natural that the history of the universe and the facts of science, interpreted in this way, should lend no countenance to the opposite theory.

The history of the universe may also be interpreted as a manifestation of the Divine will, the process of evolution as a progressive revelation; and if any be tempted to say with a sigh, "Ah! but it all requires us to believe that there is a God, to begin with," let them reflect that the other interpretation cannot even begin without the assumption that there is no God.

But to return to our chess-men. A closer study of the game would reveal--in addition to the invariable sequence of Black, White, Black--the fact that the various pieces had various properties and moved in various ways, some only one square at a time, some the whole length of the board; some diagonally, some parallel to the sides of the board. Further, our philosopher would observe that each piece when it moved tended to move according to its own laws: in the absence of counteracting causes, e.g. unless some other piece blocked the way, a bishop tended to move diagonally the whole length of the board. As a man of science, he would state these observed uniformities in the hypothetical form rightly adopted by science: if a castle moves it tends to move in such and such a way. Thus eventually he would be able to foretell, whenever any piece began to move, what direction it tended, in the absence of counteracting causes, to take. He might not, indeed, be able to say beforehand which of White's pieces would move in reply to Black, but his knowledge of the game would eventually become so scientific that he would be prepared for most contingencies, i.e. be able to say approximately where any piece would move if it did move. That knowledge could be attained without making any assumption as to whether free-will or necessity was the motive force expressed in the game; and it would be equally valid whichever of the two assumptions he chose to make. His science would have nothing to hope or fear from either assumption.

With regard to matter and motion, he would note that a piece might be removed and deposited by the side of the board, but was never destroyed, and he would infer that matter is indestructible and could never have been created. As for motion, the condition, the only invariable and necessary condition, of movement is previous movement, Black must move before White can: the only condition of change in the distribution of the pieces on the board would be some previous change. If the suggestion were made to him that possibly the real condition of all movement and every change was the purpose of an unseen agent, and that real knowledge was impossible without some idea of that purpose, he might as a man of science decline to accept the suggestion. The object of science is not to conjecture why things happen, or with what purpose, but to describe positively the way in which they actually do happen, or perhaps merely to describe the motions of material things in space. It does not matter with what purpose a shot or a mine is fired, or even whether with any or none: the results are just the same, if it is fired in just the same way. Science neither assumes nor denies the existence of purpose, because neither the assumption nor its rejection would in the least help her to discover the things that she wants to know. But are the things she wants to know the only things worth knowing? Every man is entitled to answer that question for himself. Are they the only things that can be known? They are the only things that can be known--on her assumptions. Just as the world can only be explained scientifically on the assumptions of science, so it can only be interpreted morally or religiously on the assumptions made by religion and morality. The only end that could be subserved by assuming a Divine purpose would be at most to enable us in some slight degree to argue what the purpose of some things might be--and that is of no interest or value to science. She declines to look for a final cause: her business is with efficient and mechanical causes.

The suggestion, then, that the chess-men may be moved with a purpose is not rejected, but is set aside as useless for a scientific comprehension of the game. Invisible agents--and we are all invisible, though our bodies are not--moving the chess-men with a purpose, or cross-purposes, are hypotheses valueless for science, which aims only at positive facts, the laws according to which the pieces actually do move. By the aid of these laws our philosopher might succeed in reconstructing the past history of the game which he was watching. From the positions occupied by the pieces now he might infer the positions from which they came (or think he could), and so back, step by step, until he reached the order in which the pieces are arranged at the beginning of a game. When he reviewed the knowledge thus obtained he would see in the process of the game a certain evolution from the relatively simple movements of the pawns which began the game to the highly complex movements of the queen. Then, whatever the order in which the pieces happened to be brought out and their qualities developed in the particular game he was watching, he might argue on the theory of necessity that that was the only order in which those properties could have been evolved. On the principle that efficient and mechanical causes were sufficient to provide a scientific explanation of the game it would follow that the higher powers manifested by castles and queens, the latest pieces to come out into the game, were caused by the previous action and movements of the less highly developed pawns--that life and consciousness are due to material causes. The idea that the movements of queens and pawns alike were due to the will of an unseen agent acting with purpose is, as we have said, a suggestion quite valueless to science, because any conclusions it might lead to would not be scientific knowledge. If we assumed the existence of purpose, and even could conjecture dimly its nature, we still should have made no addition to those positive facts which are the only things that science is concerned to establish: it would be neither more nor less true than before that bishops move diagonally, pawns one square at a time, gravitating bodies at the rate of sixteen feet in the first second, and so on. It would be neither more nor less true than before that pawns actually were the first pieces to move in the game, that lifeless matter preceded the evolution of organisms. Above all, it would be neither more nor less true than before that the conclusions of science are the only conclusions that a rational man will accept.

FOOTNOTES:

[37] To say that my consciousness offers no such evidence is, if true, irrelevant. We are concerned with the consciousness of mankind generally. In astronomy the personal equation is allowed for; and in science generally the observations of one savant are subject to confirmation or correction by others.

XI.

THE COMMON FAITH OF MANKIND

It is an article of the common faith of mankind that consciousness is good and trustworthy evidence of the reality of that of which we are conscious. It is also characteristic of that common faith to believe in the trustworthiness of the Power which manifests itself in that of which we are conscious. The man of science shares in the common faith of mankind up to a certain point: he accepts the testimony of consciousness to the reality of material things, and he believes that the Power which manifests itself in them can be trusted to behave when it is (in time or space) beyond the range of his observation in exactly the same way as it does within. But to walk in the common faith further than this point is unscientific. It is rational to trust the evidence of consciousness when it testifies to the reality of material things, but not when it testifies to the reality of our moral ideals, or the freedom of the will or the reality of God. It is scientific to trust the Power which manifests itself in consciousness to behave with the same uniformity in the future as it has done in the past, and rational to formulate our science and stake our material interests on that uniformity. But it is not rational or scientific to trust that Power to will freely the good of all things, or to trust our lives to that will.

The reason of this sharp division between science and faith is the mistaken idea that science involves no faith and is a body of knowledge built up without any assumption. But even if we got the man of science to admit that science would be impossible if things were not real and Nature not uniform, it would still be open to him to say that he considered any other assumptions unnecessary; and there is a way in which he could prove them to be unnecessary. He might show that they were no assumptions at all, but logical consequences from established scientific facts. That was in effect the object aimed at, as far as our moral ideals are concerned, by the optimistic philosophy of evolution.

For the optimistic philosopher, then, who refuses to begin by taking the difference between right and wrong on faith, the problem is, granted the reality of material things and the uniformity of Nature, to show that the moral law is simply one particular case of the uniformity of Nature.

The means by which this demonstration is supposed to be effected is the law of the survival of the fittest. It is shown that the law of organic life is the survival of the fittest, and that survival is the consequence of adaptation to environment. These two laws are of course uniformities of Nature. It follows, then, that there must be a constant tendency on the part of the environment to secure better and better results in the way of organic life, for it only permits the survival of the fittest and the increasingly fittest. Man is an organism, and man's good therefore consists in his adapting himself to his environment. Thus the laws of morality are shown to be but one special case of a certain uniformity of Nature, viz. the law of adaptation to environment, which applies to all organisms and not merely to man's.

The argument, however, is in the first place circular: "fittest to survive" simply means "best adapted to the environment." Doubtless the best adapted to the environment are best adapted to the environment, but it does not in the least follow that they are therefore morally or æsthetically best. There is, therefore, no such constant tendency on the part of the environment to secure moral progress as is required by the Optimistic Evolutionist.

In the next place, on its own showing, the argument ends by proving that morality--what ought to be--is nothing more or less than what is. And though that is exactly what the optimist undertook to show--and exactly what is undertaken by every one who engages to show that faith is unnecessary in morality because the laws of morality can be deduced from the facts of science--still it may be doubted whether the conclusion "whatever is, is right" is exactly either a law of morality or a uniformity of Nature.

The question at issue between science and faith is, as we have said, not whether it is possible to gain trustworthy knowledge of the world without faith, without making assumptions, for science itself is built on faith in the reality of things and the uniformity of Nature, but whether the assumptions of science are the only assumptions that we need make. One way of proving that they need not be assumed would be to show that they can be proved by science. But that way failure lies, as is shown by the optimist's ill-success. But there is yet another way of cutting down the common faith of mankind to the narrower creed of science, and that is to show that the remaining articles of faith, the assumptions not necessary to science, are inconsistent with science. That is the method adopted by the Pessimistic Evolutionist. He does, indeed, go further with the common faith than the optimist did. Impressed by the failure of the optimist to exhibit the laws of morality as the mere outcome of the laws of Nature, and the reality of our moral ideals as derived from the reality of material things, he accepts the common faith of mankind in the law of morality as being just as rational as his and their faith in the uniformity of Nature. But having taken this one step, having adopted this additional article of faith on faith, he refuses to go any further. He accepts without evidence the assumption that there are certain things which we ought to do, just as he accepts without evidence the assumption that Nature is uniform. But he refuses to accept the assumption that will is free, because that is opposed to the evidence. He admits that we ought to choose certain things, but denies that we can choose them; and his forecast of the future is in accordance with the premises from which it is inferred. It is a pessimistic picture of man being steadily driven to do the things that he ought not, ending with the triumph of what must be over what ought to be, of physical necessity over the morally right.

The object of science is to discover what we ought to believe, to substitute reasoned knowledge for ignorant conjecture; and the fundamental faith of science is that we ought not to believe anything that is contrary to the uniformity of Nature. Nothing ought to shake our faith in that article of our creed: no amount of evidence will convince a really scientific man, a true believer in the faith, that any alleged violation of the uniformity of Nature can be real. No amount of evidence would be sufficient, for instance, to warrant the belief in miracles. Either the alleged violation is only apparent, and will, with further knowledge, turn out to be a fresh instance of the truth that Nature is uniform; or else the evidence will prove on examination to be untrustworthy. To admit that any evidence could suffice for such a purpose would be to admit that the uniformity of Nature is not the fundamental reality in the world of science, or the ultimate base of our knowledge of what does actually take place in Nature.

A little reflection is enough to show that this is an entirely self-consistent line to take up. No amount of evidence can shake what is itself built on no evidence. If the belief in the uniformity of Nature depended on evidence in its favour, then evidence against it might overthrow it. But, as it rests on faith, it is superior to evidence.

Now, what is true and self-consistent in the case of science in its own sphere is equally so in the case of morality. It is the common belief of mankind that we can, and are able to, choose what is right; and just as no amount of evidence will convince a really scientific mind that a violation of the uniformity of Nature is possible, so there is no evidence which will convince a really moral man that he could not have done right when he did do what was wrong. "We ought, therefore we can," does not exactly express the facts. Rather, it is the other way: we can love, be merciful, tender, compassionate, therefore we ought. Liberty itself is a law to the free, the source of moral obligation, the gift of Him "whose service is perfect freedom."

The pessimist, then, who thinks, by producing evidence, to show that what ought to be cannot be, is adopting in morality a form of argument which in science, when it is a question of miracles, he condemns as inherently vicious and illogical. Further, the evidence which he does produce is not altogether above suspicion. It takes the form of the statement that the uniformity of Nature is a uniformity of necessity and not of a will freely purposing a good end by means of a voluntary uniformity.

If that statement could be proved to be a logical consequence from the facts of science, then it would indeed be proved that one article in the common creed of mankind was inconsistent with the rest. But, as we have argued already, it is not implied either in the admitted uniformity of Nature or in any of the facts deducible from it. To revert to the simile of the chess-board, it is as though one should say that because Black could not have moved his knight unless White had moved his pawn, therefore White was bound to move the pawn.

We cannot, therefore, consider that the pessimist has succeeded in showing that the articles of the common faith which he accepts require in their logical consequences the rejection of the rest. In saying that man ought to choose the right, but has no choice between right and wrong, he is not formulating a consequence of the facts of science, he is simply assuming without evidence the existence of a universal necessity of which the changes in Nature and the actions of man are but the varying though inevitable expression--an assumption which invalidates morality without adding to the truth of science.

There are those whose belief in demonology furnishes them with a reason and an excuse for the misdeeds of man. The belief in necessity exhibits demonology as a doctrine of science: man would fain do right, but the uniformity, which is the necessity, of Nature allows him no choice. It is Nature, the environment, which is the abode and headquarters of necessity, the enemy of the ethical process, the arch-demon of scientific demonology. And the proof that he exists is that he must. What must be, must be, because it must.

The attempt to render morality scientific ends in a result fatal to morality; and the reason seems clear. It is that science is not morality, nor are the principles of science those of morality. Science is knowledge, morality is action. Knowledge, to be knowledge, has to presuppose that Nature is uniform and that the things it deals with are real. So, too, action, to be moral, requires the belief that our moral ideals are real and that we are free to choose between good and evil. The optimist who would have us believe that science includes all the remaining articles of the common faith, and the pessimist who argues that it excludes them, alike fall into the error of imagining that science, the knowledge of what is, is the whole of knowledge, and that the assumptions which are required in order to describe what is will enable us to do and to know what ought to be. Science, which is a true description of part of our experience, becomes a misleading half-truth when it is offered as an exhaustive account of the whole. If, knowing the rules of chess and having a record of the moves in a solitary (and unfinished) game, we refused to inquire why the pieces moved, on the ground that if we succeeded in the inquiry we should have made no addition to our knowledge of the way in which the pieces do move, we should never understand the game. But we should be nearer the truth than if we assumed that a piece caused its own movements or those of the other pieces; and that will or purpose was quite incompatible with the uniformity of their movements.

The fact is that we have to play the game--we are not merely spectators--and as a matter of fact, also, men do assume that they can freely choose what moves they will make and that there are certain moves which they ought to make. The assumptions which they make, not exactly for the sake of playing the game, but in the act of playing it, are neither included in the assumptions of science nor excluded by them. To play the game at all, it is necessary to have some knowledge (or to act as though we had some knowledge) of how the pieces move, to know that bishops move diagonally, that bodies tend to gravitate at a certain rate. Man cannot indeed act or make the slightest movement without deflecting or starting some of the processes of Nature and of his own psychological mechanism: it is through them that he operates, and by means of them that he plays the game. In the beginning he has but little knowledge of what the consequences will be if he touches this or that spring of the mechanism. Yet the knowledge is necessary for him, if he is to play the game as he ought, i.e. to attain the moral ideals of which he is more or less (less at first) conscious. In acquiring this knowledge he uses his faculty of abstraction, that is his power of concentrating his attention on one aspect of a thing or problem, to the exclusion of the rest, in order to gain a clearer knowledge of it by giving it his undivided and undistracted attention. Thus, in order to understand how the mechanism of Nature or human nature actually does act, he concentrates his attention on the working of that mechanism in the abstract, i.e. wholly apart from the fact that it is at times started, at times interrupted, or redirected for the sake of realising (or thwarting) his ideals. The knowledge thus gained is science, and is, according to the agnostic, the optimist, and the pessimist, the only knowledge that man can have.

But it is clear that man can and does reflect, not only on the way in which the mechanism acts, but also on the use to which he puts it and the relation of that use to his ideals. These reflections may add nothing to his science, to his knowledge that rooks when moved must be moved parallel to the sides of the board, but they do add to his knowledge of the game. In fine, man gains a more important part of that knowledge by or in playing the game than he does by studying the rules. The rules acquaint him with the resources which are at his disposal, the capacities of the various pieces and the powers of the various forces of Nature or human nature. But it would be absurd to pass this off as a complete knowledge of the game. We may, by playing the game, add only to our knowledge of how the game ought to be played, of how the mechanism of Nature and human nature ought to be used, and not add to our knowledge of the fact that if and when the mechanism is set agoing it acts in the way described by science. But the one kind of knowledge, though not science, is just as true as the other, on the same terms, viz. if you accept the assumptions presupposed by it.

Science, then, is from the very terms of its constitution abstract, i.e. essentially incomplete. The very terms on which alone science is possible are that it shall study one aspect only of Nature, the mechanical, and shall ascertain what conclusions follow if we confine our attention to the mechanical factors and neglect certain other factors--the freedom of the will, final causes, and the moral and æsthetic ideals--which, though voluntarily neglected for the moment, are yet known to be important factors in the game of life as it is played by us. As often as we act, however, we set those factors, temporarily neglected by science, in action; and there is no reason why, when we have acted, we should not reflect upon our action, disengage the assumptions which are presupposed by our actions, and then reconsider the world and life in the light of the assumptions on which our actions and the actions of all men are based, viz. the freedom of the will and the reality of our ideals. Thus viewed, the world becomes the scene and life the opportunity of using the forces of Nature and our own psychological mechanism for the purpose of achieving the ideal.

But free-will and the moral ideal are not the only factors in the world as it is presented to the common consciousness, or in life as it is carried on by humanity, which are neglected by science, and which have to be restored by subsequent reflection, if we wish to see life true and see it whole. Science declines to entertain the question why things happen, or whether there is any purpose in events; and moral faith only guarantees that there is that which man ought to do, and that he is free to do it. But science, in neglecting the action of final causes, omits a factor which not only must be replaced before we can have any adequate understanding of the part which man plays in the world, but which, by the testimony of the common consciousness of mankind, manifests itself in the phenomena of Nature.

The description which science gives of the sequences and co-existences of material and physical phenomena is consistent with itself, and is all that is required by the assumptions of science. It is only when we reflect upon the further assumptions which we make, or rather act upon as moral agents, that we find science inadequate or--if it professes to be the whole account of the world and man--misleading. And it is only by restoring those factors for which our moral consciousness is the evidence that we can remedy the defect or correct the error. The attempt made by the optimist to dispense with the testimony of consciousness to the reality of the moral law, and the attempt of the pessimist to dispense with the freedom of the will, were both failures.

In the same way, both the scientific and the moral interpretation of the world are judged by the religious consciousness to be abstract, and are seen, when viewed in the light of its presuppositions, to be inadequate, if not misleading. The inadequacy of the moral assumptions which are made by the common consciousness of mankind is manifest, when we reflect that while those assumptions serve to decide the question--left open by science--as to the "Why?" of human actions, they do not decide the same question as to the events of Nature, but leave it open as it was left by science, whether final or mechanical causation is the ultimate explanation of Nature.

The problem what we are to do and to think in life and of life is one which for its solution requires that the whole of our experience should be taken into account: if it is to account for the sum total of the facts of which we are conscious, it must take for its basis the totality of those facts and nothing less extensive. It is true that the very vastness of the field to be surveyed--the whole of the common consciousness of mankind--makes a division of labour necessary, and compels us to concentrate ourselves at different times on different aspects of it, and to treat each of the phases of our experience--religious, moral, and scientific experience--for the moment as though it alone existed. But it is equally true that this isolation of first one phase and then the other is merely a temporary device, designed and adopted for a purpose; and that that purpose is to enable us ultimately to bring the whole of our experience to bear on the problem of what to do and to think.

Legitimate as it is, when we are working at the details of the problem, to distinguish the moral consciousness from the scientific, and the religious consciousness from the moral, it is necessary to bear in mind that these distinctions are merely abstractions. In thought we may and do so distinguish, but in fact and experience consciousness is a unity. The same man who is conscious of sense-phenomena is also conscious of moral obligation: the "I" which is conscious of moral experience is the same "I" that is conscious of spiritual experience.

Further, the evidence which we have for the three kinds of experience--scientific, moral, and spiritual--is the same: it is the evidence of consciousness--the only evidence that we can have of anything. That witness, if discredited at all, is discredited for all in all. If discredited, it must be by its own testimony, for we have no other witness which can give evidence against it. But we hope that it is true: the man of science is so certain of its truth, in the department in which he is most familiar with it and has the best right to speak of it, that he lays it down as a rule that there simply can be no evidence of an exception to the uniformity of Nature. The moralist is equally certain that no exception to the law of moral obligation is possible; the religious mind that there can be none to the universality of the Divine love. To the unity of consciousness corresponds the unity of our faith in its trustworthiness. Scientific and moral faith are not different from religious faith; they are but phases of the same. The common faith of mankind is not a synthesis formed artificially by adding the three together; on the contrary, the three are artificially distinguished by thought--they do not correspond to fact, but are abstractions from the facts, and are formed by the suppression of facts.

The religious consciousness is itself abstract; and as an abstraction, i.e. if taken to be the whole of what we know and feel and do, is capable of leading to false conclusions: no religious belief can stand permanently which runs counter to the facts of science or the moral faith of mankind. No amount of spiritual experience will add to our knowledge of chemistry or physics, or be valid evidence against any truth of science. It may serve to prevent the premature acceptance of something too hastily put forward as a scientific fact, in the same way that science may overthrow a belief erroneously supposed to be religious.

But though the religious consciousness is an abstraction, in the same sense that the scientific and moral consciousness are abstractions, each is valid in its own sphere; and the whole evidence of consciousness in all its three phases must be taken together, if we are to elicit any universal principles of thought and action, any unity in our experience, any purpose in evolution. From this point of view we shall expect to find a unity of experience corresponding to the unity of consciousness, and to discover that there is a fundamental identity underlying the apparent diversity in that reality of which in consciousness we are aware. What gives this unity to experience is the permanence which we attribute to the real, in whatever way the real is apprehended: the real, whether apprehended in sense-experience or in moral conviction or in spiritual experience, is characterised by permanence, as distinguished from the passing feelings with which we view it and from the transient experience we have of it. The reality of the things of which we are aware through our senses is conceived as something permanent, and is implied to be so conceived by all theories of evolution which wish to be taken seriously. The permanence of moral obligation is not conceived by those who are genuinely convinced of its reality to vary or to come and go with the flickering gleams of our moral resolutions. Nor when spiritual light is withdrawn from our hearts is it supposed, by those who believe it to be the light of God's countenance, to be quenched for the time.

The fundamental identity of the real throughout its diversity is what is postulated by science when it explains the process of evolution by means of the law of continuity. It is equally postulated by the moral philosopher who claims objective validity for the moral law on the ground that it is the same for all rational minds. It is the faith of the religious mind which not only feels the Divine love in its own heart, and finds it every time it obeys the conscience, but also divines it in the uniformity of Nature and throughout the process of evolution.

The identity of the real does not lie in the mere fact that we are conscious of it. The real things of which we are conscious have, indeed, as one feature common to them all, the fact that we are conscious of them. But the identity of the real is not created by nor a mere expression of the unity of our consciousness. It is not the understanding which makes Nature--save in the purely psychological way in which apperception does; on the contrary, the things of which we are conscious in sense-perception are given as independent of us, though sense-phenomena are obviously not. In the same way, the reality of the moral law is conceived, in the very act by which we recognise it as binding on us, to be something independent of us; nor is God's love towards us dependent on our merits, or existent only when we recognise it.

If, then, we are to gather up the permanence, the identity, and the independence of the real into the unity of a single principle, if we are to interpret the law of continuity in the light of the whole of our experience, we must look to the Divine will. In it we shall find the reality which is progressively revealed in the law of continuity; in it we shall find the permanence and the independence without which reality has no meaning; in it the changeless and eternal identity of Him whose property it is ever to have mercy and always to be the same. Then, perhaps, we may extend the principle of scientific method so as to include the whole of our experience and to make the whole of our knowledge truly scientific; for to the uniformity of Nature and of human nature we shall add the uniformity of the Divine nature, or, rather, we shall see in the former the expression of the latter. But it is not the agnostic who will thus enlarge the bounds of science, or open a page of knowledge rich with the spoils of faith.

XII.

PROGRESS

The artificial nature of the abstraction which distinguishes the scientific from the moral and the religious consciousness, as well as the impossibility of simultaneously exercising faith and repressing it, is plainly exhibited in the optimistic interpretation of evolution. The premises from which it starts are faith in the uniformity of Nature and belief in the reality of material things. The conclusions which it reaches constitute a non sequitur if they are supposed to follow from the avowed premises, and only command assent when we tacitly assume certain moral and religious presuppositions which, if not avowed in the optimist's argument, are instinctively supplied by the moral and religious consciousness of the optimist's disciples. That the process of evolution on the whole has been and will be a process of progress follows logically enough from the optimist's avowed premises, if by progress we mean the survival of those best fitted to survive--that is, if we empty the notion of progress of all moral meaning. But as the conclusion that evolution is progress is the conclusion which is necessary for the justification of the common faith of mankind, the illogical nature of the optimist's process of inference is apt to be overlooked in consideration of the satisfactory termination of his argument.

It is, however, necessary, in the interests of clearness of thought as well as of the moral and religious consciousness, that the conception of progress thus thoughtlessly emptied of meaning by the optimist should have its context restored. This service--a service essential as a preliminary to every theory of evolution--was rendered by one in whom the moral consciousness spoke with force--Professor Huxley. To him is due the demonstration that adaptation to environment, so far from being the cause of progress, counteracts it; so far from being man's ideal, it is that which resists the realisation of his ideals. Progress is effected, according to Professor Huxley, not by adaptation to but adaptation of the environment, and consists in approximating to the ideals of art and morality--which ideals are not accounted for, as ideals, by the fact that they are the outcome of evolution, because evil has been evolved as well as good. Why approximation to the ideal of religion--love of God as well as of one's neighbour--should not contribute to progress does not appear. If, however, we add it, and also add the ideal of science, viz. truth, then progress will be the continuous approximation to the ideals of truth, beauty, goodness, and holiness; and human evolution, so far as evolution is progress, will be the progressive revelation of the ideal in and to man.

Two things are implied in this conception of evolution: the first is that evolution may or may not in any given case be progress; the next that we have a means of judging, a canon whereby to determine, whether evolution is progress. Both points are illustrated by the argument of Professor Huxley, who uses the moral ideals as a test whereby to judge the process of evolution, and decides that evolution has been progressive in the past and will be regressive in the future. Strange to say, the reason why Professor Huxley maintains that evolution will be regressive is exactly the same reason that leads Mr. Herbert Spencer to maintain that it will be progressive. It is that the law of evolution is Necessity, that evolution is the outcome of mechanical causes. But in effect both arguments lead logically to the same conclusion, for the progress which is the outcome of Mr. Spencer's argument is not progress in the moral or any other sense of the word. In fine, progress is eventually impossible if evolution is due to mechanical causes; progress is conceivable only if we interpret the process of evolution teleologically and as expressing the operation of a final cause. Science, as such, declines to inquire whether there is any purpose in evolution, and leaves it an open question. The moral consciousness affirms only that the process of evolution ought to make for good. The religious consciousness alone is in a position to say that its spiritual experience requires us to affirm that evolution, in accordance with the uniformity of the Divine nature, will be, in years to come as in ages past, a continuous movement towards the realisation of all that in its best moments the human heart holds most dear.

The argument that evolution must be progress commits logical suicide, for in the very act of proving its conclusion it proves that progress is not progress. We are therefore left to face the fact that progress is only a possibility; and that amounts to saying that regress also is possible. What is implied therein will become clear if we return to the question of the nature of progress.

Progress is not the survival of the fittest to survive, but of the aesthetically or ethically fittest; not adaptation to the environment, but approximation to the ideals of truth, beauty, and goodness. Those ideals are manifested in man, but not equally in all men; and the words and works of those men on whom they are most clearly impressed and by whom they are most faithfully expressed become the canon whereby we judge whether any tendency in art or morality is progressive or retrogressive. We cannot all make beautiful things or do heroic deeds, but we can all appreciate them when made or done. To appreciate them, however, is to judge that they do come nearer to the ideal than anything else of the kind which we have yet known. Thus the ultimate court of appeal for each one of us is not the ideal as manifested by man, but the ideal as revealed to each of us. True it is that, until we saw that particular work of art or that particular instance of love, we had no idea what beauty or love could be. But that makes no difference to the fact that we feel for ourselves how much nearer it comes to the ideal than anything we had any idea of before. It may henceforth be the standard by which we shall measure other things, but in adopting it we measure it ourselves, and measure it not by itself, but in relation to the ideal. And what shall we say of the artist himself? By what does he measure the work of his predecessors and judge that it is not the best that can yet be done, if he does not measure it by the ideal which is revealed more perfectly to him than to them?

But the perfect work of art or love, when done, becomes not merely the canon by which to test progress; it becomes itself the cause of progress, both because of its more perfect revelation of the ideal and because of the emulation which it arouses in others to go and do likewise. They likewise strive after the ideal and labour for its sake: it is the final cause of their endeavours, the purpose of their attempts; and were there no such final cause there would be no progress. The ideal is a principle both of thought and action, the test of knowledge and the source of progress. Truth is the ideal of science: approximation to truth is that for which the man of science labours and that in which he conceives that scientific progress lies. The gravitation formula not only expresses a wide-reaching truth, but has acted as an incentive to many attempts to extend it to the domain of chemistry, and serves as an ideal yet to be rivalled in other branches of science. But science and progress in science are alike impossible, if consciousness and experience be discredited, or if the ideal of science be not real, i.e. if the laws of science have not the permanence, the independence, and the self-identity which are the attributes of the real, but are as transient as the minds that discovered them, exist only when thought of by man, and are not really the same at different times. But if these ideals, whether of truth, beauty, or goodness, are thus real, then they are "our" ideals only in the sense that we are aware of them and adopt them, not in the sense that we make them; they are ours because they are present in the common consciousness of mankind, but not in the sense that they are created by that consciousness. They are revealed to man before they are manifested by man.

Professor Huxley's definition of progress cuts at the root of two misconceptions as to its nature, which, though mutually inconsistent, are both widely spread. One is that the latest products of time, simply because they are the latest, are superior to all that has preceded. The other is that to know the origins of a thing will best enable us to assign its value. The tendency of the one is to result in the idea that because a thing has been evolved it must be superior; of the other to lead to the conclusion that because a thing has been evolved out of certain elements it cannot be superior to them. The truth is that the mere fact that a thing has been evolved--be it an institution, a mode of life, or a disease--does not in itself prove either that the thing is or is not an advance on that out of which it was evolved. Regressive metamorphosis, degeneration, pathological developments--physiological, mental, moral, and religious--are all processes of evolution, but are not progress. A society in its decay, or an art in its decline, is evolved out of a previous healthier state or more flourishing period, but is not because later therefore better. Nor, on the other hand, does it follow, because the earliest manifestations of a tendency are the lowest, and can be shown by the theory of evolution to be so, that no progress has been made in the process of evolution. The artistic impulse in its earliest manifestations, in children and in savages, is rude enough; but it would be absurd to say, therefore, that art in its perfection has no more value than in its origins, that the Hermes of Praxiteles is on a level with a misshapen idol from the South Sea islands.

If the continuity of evolution does not warrant us in assigning the same value, aesthetic or moral, to all the links, highest and lowest, in the chain, still less does it authorise or require us to deny all value to the lowest. On the contrary, we should rather see in the lowest what it has of the highest, than look in the highest for the lowest we can find. We should beware lest in reducing everything to its lowest terms we prove to have been seeking simply to bring it to our own level, when at the cost of a little more generosity we might have raised ourselves somewhat nearer to the ideal prefigured even in the lowest stage of the evolution of love, of beauty, of piety or of goodness. Indeed, as a mere matter of logic, it is impossible to state the nature of a cause accurately, quite apart from any question of estimating its value, until or unless we know the effect which it produces. It is not only that we may underrate or entirely overlook the importance of a thing, so long as we are ignorant that it is a factor largely influencing some result in which we are interested; but, until we know what effects it is capable of producing, we do not know what the thing is. We could not be said to have knowledge of a drug if we did not know what its effects were. Nor is that knowledge to be acquired by analysing the causes which produce the drug. It is not from the mechanical causes which give rise to a thing that we can learn what a thing is: no amount of knowledge of the properties of hydrogen and oxygen would enable us to predict a priori the nature of the compound which is formed when electricity is passed through two molecules of the former and one of the latter; nor is the least light thrown upon the properties of water by our knowledge of its constituent elements: on the other hand, our knowledge of them is materially and serviceably increased when we learn what they are capable of producing in combination. We learn most truly what a thing is from observing what it becomes, what use it subserves, what end it answers, what purpose it fulfils--in a word, when we know not its mechanical but its final cause. In biology, knowledge of an organ means knowledge of its function--that is, of its purpose; and evolutional biology also teaches that function is the cause of the organ.

It is by observing what a thing becomes that we learn the part it may hereafter play in the general scheme of things, and come to know its real nature, and estimate it at its real value. Thus our estimate of the value of such an institution as "taboo" goes up, and our knowledge of it is increased, when we recognise in it one of the early manifestations of the sense of moral obligation on its negative side. Again, in tracing the evolution of religion it is impossible to know which of the various rites and ceremonies, practised by a savage tribe in its dealings with the supernatural, are religious and which non-religious, without taking into consideration the question, What do such customs tend to develop into? Until we know that, and until we can say whether what is evolved out of them is religious or non-religious--a question which we cannot answer unless we know what religion is--we cannot be said to understand the nature of the savage rites that we are studying. But it is not from the origins of art, religion, or morality that we shall gain the answer to the question what art, morality, or religion is; for the question must be answered before we can recognise the origins when we see them, and can only be answered by reference to the ideal, which is the test and final cause not only of progress, but of the real.

The ideal is a principle both of thought and of action. As a principle of thought it is the test by which we determine whether any given movement is progressive or regressive, and whether any given thing is what it appears or is alleged to be. As a principle of action it is that for which we strive, the purpose with which we act, the cause of any progress that we make. If we are not prepared to maintain that everything which takes place is an advance upon what preceded, we require some test whereby to distinguish what is progress from what is not, and we admit that progress is a possibility which may or may not be realised, and it becomes of interest to inquire on what conditions its realisation depends.

If, as Professor Huxley maintains, the test of progress is approximation to the ideal, then one condition of progress is that man shall be conscious, to whatever extent is necessary for the purpose, of the ideal, shall feel that the ideal of love, tenderness, compassion, justice, truth, beauty, etc., is a thing for him to strive for, an end for him to attain. To the chosen few--the great artists, the moral or religious reformer--a sense of the ideal is dealt in a larger measure than to the rest of men. By the chosen few it is manifested to the many. But it does not become the cause of progress, unless it leavens the mass, unless they too are inspired by it to do better and be better. In a word, when, or if, ever the ideal has been manifested in its fulness, it is not a fresh revelation which is necessary for progress, but fresh conviction in us and renewed determination. Indeed, so long as we do not act up to the light that we have, even an imperfect revelation of the ideal may serve for imperfect beings.

So far, then, as the genius in art or science, or the reformer in religion or morals, is the cause of the progress that is made by his school, his disciples, and them that follow after, it is clear that he is the cause and not the product of evolution. It is his works or words which inspire his followers with a fresh sense of the reality of the ideal and a fresh resolve to devote their lives to the pursuit of art or the service of science. But it is only because his perfect work is felt by them, judging for themselves, to realise the ideal that it has this effect on them; and they could not judge it to approach the ideal more closely than anything known to them before, unless they had some surmise, however vague, of the ideal, with which to compare this work, perfect as it seems to them. It is not necessary to suppose that this vague surmise existed, or if it existed that it was attended to, previously: it may have been first called into existence or into notice by the contemplation of the master's work, but its presence, however evoked, is attested by the judgment that his work does come nearest to the ideal. The manifestation of the masterpiece may be the occasion of this fresh revelation of the ideal, but the revelation must be made if the work is to be judged highest and is to inspire the disciple.

It is, however, one thing to have an ideal, and another to live up to it. "To scorn delights and live laborious days" in the search for truth or in single-minded devotion to the cause of art requires some will. Granted that the ideal has been revealed, either to the disciple on the occasion of another's teaching or directly as to the master, for progress there is further required will. It requires an act of will to prefer the ideal, with its laborious days, and to scorn delights; and it requires many acts of will to make any progress. Yet the will to believe and the will to act are the same will. We may, if we choose, define belief as the readiness to act, and take action as the test of belief: if a man in a hurry makes a short cut, i.e. goes straight from one point to another rather than round a corner, his action is proof that he believes that a straight line is the shortest distance between two points. From this point of view we may regard the many acts of will which are necessary to progress, i.e. movement in the direction of the ideal, as so many reaffirmations of the original act of will by which we affirmed our belief that the ideal was the goal of progress; and if our object is to show that the behaviour of man, so far as he pursues the ideal, can be exhibited as a logical and rational behaviour, we are justified in thus demonstrating that our renewed resolutions to realise the ideal are but the logical consequences of our original will to believe in the ideal as the proper goal of action. Our belief in the ideal is thus shown to be the principle from which our subsequent acts of will can be logically deduced, just as Nature's uniformity can be shown to be the principle from which the conclusions of science logically flow.

But it may be doubted whether this logical order of ideas is the chronological order of events. As a matter of fact, we go through a number of struggles and temptations long before we reflect, if ever we do reflect, upon them in such a way as to see what is the general principle logically implied by our repeated if intermittent resistance to temptation, just as a child acts in a way that for its logical justification would require a formal recognition of the uniformity of Nature, though the infant of two years, or less, does not formulate that principle as a condition precedent of crying for its food or its nurse. Chronologically, then, the will to act seems to precede the will to believe in the uniformity of Nature, and in the case of most human beings is never followed by any fully conscious formulation of the principle on which we act as an abstract principle in which to believe. That fact, however, does not in the least detract from the value which the formulation of the abstract principle has: when formulated it becomes in the hands of science as Ithuriel's spear for the detection of lingering superstitions and confusions of thought--

"for no falsehood can endure Touch of celestial temper, but returns Of force to its own likeness."

At touch of the question, "Does it contradict the uniformity of Nature?" error is seen for what it is, and is exploded sooner thus than in any other way.

The ideal of truth, then, with its "celestial temper," is logically implicit in the earliest acts of will, but chronologically is developed in consciousness later, if indeed and when it reaches that later stage of its evolution from the potential to the actual. The ideals of morality and religion, again, though equally implicit in the acts of will which form their earliest manifestation, are, as a rule, both in the individual and the race, more slowly evolved from the particulars in which they are immersed. The period of their gestation is longer, and results in the birth of a higher organism.

Thus, when we reach the age of reflection, whenever it may come, we wake up to find that we have been acting as though we had beliefs, when, as in our infancy, we could have had no beliefs, and as though we willed our actions, at a time when we can scarcely be said to have had any will in the matter. For years we have been acting as we should have done supposing that we had believed certain things and had willed our action accordingly. When we wake up to this state of things the question is, Are we bound to go on in this way? are we bound now to believe as well as to act as though we believed in God, morality, and Nature's uniformity? Does the fact that our physiological and psychological mechanism has been started--perhaps by Nature's cosmic forces, perhaps by the social environment, certainly not by us--to run in certain grooves, prove either that we ought or that we must continue to run the particular organism we are in charge of on the same lines? The agnostic and the atheist exercise their freedom of will to say No. They claim the right and exercise the power of free choice. The agnostic, further, is fully aware that in choosing to believe the uniformity of Nature his choice is not determined by evidence--it is "a great act of faith," no amount of evidence could justify it, the only evidence anyone can bring to justify his belief in the general abstract principle is the fact that he does believe it in every concrete, particular instance. In a word, he believes it because he chooses to believe it--and that is exactly what is meant by the dictum, which he finds it so hard to understand, that his will is self-determining.

When it comes to the question of morality and religion, the agnostic again exercises his freedom of choice: he wills to believe in the former and not in the latter--the evidence for and against either being equally nil. It is not, therefore, the evidence which determines his choice; and he shows that it is not his previous history, not the momentum which his psychological mechanism gained during the period when he had no conscious or no self-conscious control over it, which determines his choice, for in the first place he denies that it ought or must influence him, and next he shows that it does not, by willing differently in the case of the two principles. In both cases his will is equally self-determining, though his will is to believe in the moral principle or ideal and not to believe in the religious.

If we wish either to define progress or to make it, we must choose, arbitrarily or otherwise, some particular goal and say, definitely and decidedly, any movement which being continued in the same straight line leads to that goal is progress, every other movement is regress, being necessarily away from the goal. If we choose, by a great act of faith or otherwise, to say the ideal is the goal, then we have therein a principle both of belief and action: we have a standard by which to test everything offered for judgment, a general principle to apply to every particular case; and we have an object to aim at, a principle to carry out in every act of our lives, an ideal to strive for. But whether we choose the ideal as the goal or something else, our choice is the free act of a self-determining will. Progress, on the human side, is--as indeed is regress--the expression of the free will of human beings, whose choice, though free, is limited to the alternatives offered to them. Those alternatives reduce themselves ultimately to aiming at the ideal or at something else.

What, then, of the environment, of the cosmos, in which man finds himself, in which he has to act and may act so as to advance or not to advance towards the ideal? To begin with, we may distinguish between those forces in the cosmos which man can to some extent control, and those over which he has no control. The former, from this point of view, the point of view of action, are means whereby man secures his ends: his regulation of them effects that adaptation of the environment which, according to Professor Huxley, is essential to ethical progress. Now, as a matter of observed fact, no one doubts that the advance which civilised man has made in controlling the forces of Nature is due to science and to civilised man's devotion to the scientific ideal of truth. Even the savage made what little progress he did make in this direction by acting fitfully and unconsciously, or at the most semi-consciously, on the principle of the uniformity of Nature: the savage was faithful in little things to the scientific ideal--which was revealed to him but dimly--the savant is fully conscious of the principle on which he acts, walks in its light, and strives by example and precept to save his fellow-men from relapsing into the darkness of error and superstition. It is not merely because of the material advantages, the comforts and luxuries, which science indirectly secures to mankind, that the man of science devotes himself to the scientific ideal and seeks to make it universal: it is for the sacred cause of truth. In a word, what at first sight presents itself merely as a principle of the scientific reason, proves, in the conception of those who have spent their lives in endeavouring to seek the scientific ideal and to ensue it, to be a manifestation of the moral reason, to be not merely in harmony with the moral ideal, but to have been its harbinger, making the way straight for it. Belief which implies a violation of the uniformity of Nature is denounced not because it violates a scientific principle, but because it is immoral, a pretence, and a lie. The final cause of science is thus made out to be to subserve the moral ideal, to secure that adaptation of the environment without which ethical progress is impossible. The labour of adapting his environment would have for man as a rational being no sufficient reason if it did not tend to realise his moral ideal. Man may use his science and the power of adapting his environment for other than moral ends; but such use is not, according to this view, progress. In other words, it is not science or the scientific ideal alone which enables us to lay down the line of progress, but science and morality together: one point cannot give us our direction, but the line which connects two points may.

Thus far, then, by taking the environment into consideration, we seem to have introduced no new factor into our conception of progress. It seems that when I wake up from childhood's slumber I find myself surrounded by men who believe that they can do certain things--make rain, send telegraphic messages, etc.; and I am told that if certain assumptions--that there is a God, that Nature is uniform, etc.--be true, then it will be well for me to behave in a certain way. But what if the assumptions be not true? My elders tell me that experience--in the individual, in the race, enlarged by science and the theory of evolution--shows it is quite safe to assume that they are true, at any rate as a provisional hypothesis. Of course, if the future is going to resemble the past, then experience of the past is a good guide to the future: but that is just the question, is the future going to resemble the past? In other words, what attitude am I to assume towards my environment, the cosmos? Am I to assume that it will work, and for countless ages has worked, in such a way as to make it possible for me, with some co-operation on my part, to do things which my elders tell me are desirable and which I feel for myself I should rather like to do?

If I assume that the cosmic power does work thus, in such a way that I can know the truth and do the right, and love the Power that gives me the chance and makes it possible, even for me, so to do, I am only exercising the will to believe in that principle which is logically implied by every act of the scientific or moral life.

It is the common faith of mankind that experience may be trusted; and it is the common experience of mankind that progress is approximation to the ideals of truth, of goodness, and of love. It is not the common experience of mankind that all men or all peoples approximate equally to those ideals. The measure of progress is to be found in the clearness and consistency with which men have carried out in science the principle of the uniformity of Nature, in their dealings with their fellow-men the principles of morality, in their dealings with the supernatural the principle of love.

Science, and especially the theory of evolution, has enormously extended our inferential experience, but it has done so only by accepting the common faith that experience may be trusted, that is to say, that the environment, the cosmos, is trustworthy within our experience of it. When, then, the optimist alleges that the process of evolution has been, on the whole, a course of progress, he is but showing that the common faith in the trustworthiness of the reality in which we move and have our being justifies itself. But he does not show us, nor does science show us, why the real, the cosmos, is trustworthy: he ends by showing that it is trustworthy because he began, like all of us, by trusting it. He is quite right: it is the only way in which to demonstrate that either science, or morality, or religion is trustworthy--by giving our faith, to start with. Only when we are satisfied as to the fact can we profitably inquire the reason; and the reason is to be found only in the nature of the real, as revealed to us in the sum total of our experience, scientific, moral, and religious. But the will to believe that experience and to trust the real which it reveals, is free: if a man will not accept it as trustworthy, there is for him no reason why.

The case is different with the man who does accept the testimony of consciousness as evidence of the reality to which it testifies. For him the one reality is Will, and the ideals of science, morality, and religion are the expressions of that Will. In accepting them as the principles of thought and action he does not learn what is the purpose of evolution, the final cause of the cosmos: he chooses to believe that, by so accepting them and by striving to realise the ideal, he is fulfilling the Divine Will and contributing his share to the realisation of the rational purpose to which, he assumes, the process of evolution is tending.

But in so doing he does not renounce his freedom: his resolution to believe is an exercise of his free will, an act, "a great act," of faith. If carried into effect in his daily life, his resolution, daily renewed and ever free, may in the end become a daily act of love, and then he will understand the reason why the cosmos, or the cosmic power, is trustworthy. Only love of man could have given man, as his ideals, to know the truth and do the right. Only if man's ideals are so given is the cosmos trustworthy--if it is trustworthy. If it is not, then there is no truth to know, no right to do, no inference can be drawn from the past to the future, for the past, even of a minute ago, may be a delusion.

But though the will to believe that the cosmos is untrustworthy cannot in practice be carried out in all its logical (or illogical) conclusions, it can be and is acted on intermittently, and such action is regress. So far as it is carried out, it is the negation of progress; if it could be carried out completely and by all men, there would be an end of progress; science, morality, and religion would be extinguished; evil would triumph over good. The history of evolution shows that, as a matter of fact, such unfaith in the reality of our ideals has been only intermittent; for the course of evolution has been, on the whole, progress. Individual experience shows that there comes a point, soon or late, at which the will, acting freely, refuses to go further with its rejection of morality: there are some things which even a bad man will not do--however oddly they may seem chosen. In theory, in philosophy, there is a point at which the will refuses to go further with its rejection of the common reason, in which all men share: there are some things which even the sceptic refuses to disbelieve, e.g. those which are necessary to his conviction that nothing can be believed.

These considerations may serve to confirm us in the belief that progress has been the law of evolution in the past and will increasingly be in the future. They should so confirm us, for they do but carry out, as far as history, individual experience, and imagination can take us, our fundamental faith in the reality of those ideals that are revealed in consciousness to all of us. Belief in the possibility of progress at all carries with it, as its logical postulate, faith in the wisdom and goodness of God. But if wisdom and goodness are the source of all reality, and if the final purpose of evolution is the realisation of the ideal--viz. love of truth, of our fellow-beings, and of God--what are we to say of evil? Is it not real? It is real, in the same sense that our pleasures and pains are real, but not in the same sense that the ideal is real. The real things which our sense-experience reveals to us are real in the sense that they are permanent, independent of us, and self-identical. The same characteristics attach to the realities revealed to us in our moral and spiritual experience. The laws of morality and the goodness of God do not come and go with our fleeting recognitions of them; they are permanent, independent of us, and are ever the same: God's goodness faileth never. The uniformity of Nature is but one expression of the uniformity of the Divine love for man: it is that which makes it possible for man to know the truth and survive in the struggle for existence. But evil is not independent of us men: it exists only so far as we will it to exist. It is not permanent: it comes and goes with our passing acts of will.

It is not self-identical, but tends to self-destruction. It is the will to believe nothing, and therefore, as action involves belief, the will to do nothing--that is, to revert to the condition of mere inert matter, as matter is conceived by the materialist to exist.

But though evil be illusive, though it is the fool who says in his own heart, "There is no God," or "Tush! He will not see it," the illusion is voluntary. It is we who deceive or sophisticate ourselves, when we will to believe that this act is not really wrong, or that our peculiar circumstances constitute a special, a highly special exception, on this occasion only, to the rules for ordinary occasions and ordinary men. And though the illusion is subjective, i.e. is not generally shared by the onlookers, and is consciously subjective (for we avoid onlookers, because they would spoil the illusion), nevertheless, subjective though it be, it is a fact in your particular subjective history, and a damning fact. If the evil that you will is confined in its range to your will, and if its existence can only be recreated by a fresh act of will in you or another, that is an argument to show that there is mercy in the scheme of things, but it does not prove that you incur no responsibility in offering yourself or another the example and the opportunity of doing wrong. We are not, and, if the will be free, we cannot be responsible for what others do; but we are responsible for what we do--for evil, if it be evil; for good, if we----but there is no pressing need to consider that contingency.

The question underlying the previous paragraph is that of our social environment and its effects. We are apt to forget that we are the social environment. If we bear the fact in mind, we shall perhaps be less inclined to seek the origin of all our misdeeds outside ourselves: we cannot shift the burden of our own wrong-doing on to the shoulders of society by any process which does not bring back at least an equivalent burden. The fact is that neither can we cause others, nor can others cause us to do evil. What we can do is to supply them with an opportunity, which, but for our action, would not indeed have existed, but which also, so far from necessitating evil action on their part, may by their free will be made the occasion for a victory over wrong. The fact, however, that they alone are responsible for their evil-doing prevents us from taking any credit for their good deeds. It is for our own acts of will that we are responsible, and it is by willing evil that we become evil. We create evil, consciously, by every wrong act of will that we perform, and then we talk of the origin of evil as a mystery, so thoroughly do we sophisticate ourselves! Why should there be evil? Why, indeed? There is no reason, no rational answer can be given, because evil is irrational--it is the will to reject the common reason or common sense or faith of mankind, in this detail or that. It is the arbitrary element, self-will, and if it could be eliminated we should have a uniformity of human nature and of human love corresponding to the uniformity of the Divine. Progress is the process of its elimination.

If we turn from the human to the pre-human period of evolution, the first immediate fact which strikes us is that there has been throughout the animal kingdom an evolution of mind, which has resulted in providing man with the psychological apparatus necessary for conceiving, and, to some extent, realising the ideal. When we reached the age of reflection, we woke up to find that our psychological mechanism had been running for some years in certain grooves. We now find that its direction can be traced back by evolution to the beginnings of animal consciousness. If, however, we believe that the evolution of mind, animal and human, has been a process of progress, we do so not on the ground that mind has been evolved, but that its evolution has been in the direction of those ideals, approximation to which is believed by us to be progress. Similarly, if the pre-animal period of the earth's evolution is shown by science to have resulted in fitting the earth to be the home of animal life, we judge that evolution to have been progress, not because it prepared the world eventually for man, but because it is seen to have been part of the process by which the ideal is in course of realisation, by which the Divine purpose is in the course of being fulfilled.

The only value that we can assign to the pre-human period of evolution is that which attaches to it as a means to an end; but though we believe that by striving after the ideals revealed to us we are labouring towards that end, and though everything that makes for the ideal contributes to the end, yet we do not know the Divine purpose, and we cannot say in what manifold other ways the pre-human period may have subserved that purpose. It is sufficient if we can trace the steps by which this one portion, the only portion known to us, of the whole design has been carried forward. This reflection is one which it is necessary to bear in mind when considering the alleged wastefulness of the process of evolution and the price at which progress has been purchased.

The theory of evolution, as a purely scientific theory, i.e. as an objective statement of what actually has taken place on the earth in the past, shows that the various species of animals which have survived were--so long as they did survive--the only species which could survive under the conditions which then prevailed; given the conditions, their survival was necessary and inevitable. There the scientific explanation of the matter ends: having shown the causes which produced the effect in question, science has explained everything that it undertook to explain. Had the conditions been different, the present state of the world, doubtless, would have been different; but being what they were they produced that which is, and there is an end of the matter--as far as it is a matter for scientific investigation.

What we are to think of the survivors--whether we are to admire them; whether we are to consider their survival an advance and an improvement; whether anything has been gained by their survival, and, if so, from what point of view the gain is a gain--are questions which science excludes, because, however answered, they do not affect the scientific fact that these species did survive, and, under the conditions, alone could survive.

But we all take it for granted and as self-evident that man is not only better adapted, under existing conditions, to survive and flourish at the cost and to the extinction of other species, but that he is better than the brute, that his survival is an advance, that his is a higher type, and that his existence realises a higher ideal than that of the brutes. We believe this not merely because we are men, and as such rate our own comforts, our own interests, our own skins as the most important things known to us, for there are things for which men sacrifice their own interests and for which they have laid down their lives. It is precisely because there are things more important than our own material and animal existence, and because they are or may be realised by man and not by the animals, by the ideal man and not by the brute man, that we consider him to be worth more than many sparrows--though they too have their value in His eyes--and man's existence to be of a higher type than theirs.

Thus, then, when science--which, if it is truly scientific, makes no distinction of value, moral or spiritual, between man and the sparrow--has explained that a given species which did survive was the only species that could have survived under the conditions, there still remains the problem, for those to whom it is a problem, Why should the species which was bound to survive also happen to be a species of a higher type? Why have the survivors always happened to be both better adapted to survive and better adapted to further the ideal which the course of evolution reveals with increasing clearness?

In fine, science explains only a part, not the whole of the effect of evolution. It concentrates its attention on one part or aspect of the effect, on the survival of the fittest, and explains very simply and satisfactorily that the environment kills off the creatures which are not fit to cope with it, while the fittest to contend with it survive. The fact that the survivors not only are best adapted to the environment, but are also best adapted to bring the whole creation one step nearer to those distant ideals in expectation of which it groaneth and travaileth, is that part of the effect which science, for scientific purposes, rightly ignores. Science does not undertake to estimate the value of the effect produced, or even to consider whether when produced it has any value.

But when the question is raised as to the cost at which the process of evolution is carried on, it becomes necessary to bring into the account the value of the result attained or to be attained. Possibly, creation that groaned in her travail may rejoice that a man-child has been born. But so much depends on what her child grows into. And he has free will. We have the power now and here to dash her expectations to the ground.

The value of a thing to me is exactly what I am prepared to give or do for it. I have no other way of estimating the value of the ideals for which creation has laboured in the past, than by asking myself how far I am prepared to go for the love of truth, of fellow-beings, and of God. If I am prepared to give everything, and then count myself the gainer, then indeed I may know that the cost of evolution has not been greater than the value of the ideal: I know the highest price, and I know the feelings of those who pay it. And they are the only persons who can judge the value of the article, for they are the only people who get it. The fact, however, that they do get it, that they get it in full, and every man according to the measure with which he metes it, contains the answer to our question. What is true now was true of earlier generations and earlier men: the value of the ideal to every man was exactly what he gave for it. It is the realisation of the ideal by me that is my reward, though my object may be its realisation by others. But it is absurd to say that their gain is my loss, or that their progress has been made at my expense.

These considerations apply only of course to those men who have sacrificed themselves for the sake of progress and the love of their fellow-man. Most men, however, do not sacrifice themselves much; and therefore they can hardly be brought out as martyrs to the cause of progress, as the millions who have perished by the wayside in the march of evolution.

It is not until we introduce the element of material progress that it becomes possible to maintain with any plausibility that there is a divergency of interests between the contributors to it, or that they who sowed have been sacrificed to us who reap. It is when we compare the shivering savage with our sheltered civilisation, primitive man's struggle for existence with civilised man's enjoyment of existence, that we begin to be anxious about the cost of evolution--that is to say, that our little faith in the value of the ideal begins to torment us. In our unreadiness to sacrifice ourselves we forget that it is possible for civilised man also to make sacrifices--perhaps the greater because he has the more to forego--and that the savage has his tribal traditions, embodying his ideal of a good man, to live up to; his tribal customs, which he may violate with self-reproach, or fulfil with satisfaction; his conceptions of the truth about man's relations to the past, the world, and the supernatural. The savage also has his ideal, which he sets above his pleasure, and for which he faces pain in many a cruel rite. Shall we say that its realisation is no reward to him? or that in realising it he does not as faithfully contribute his mite to the fulfilment of the Divine purpose as we? We make too much of our superiority. We make, also, too little of the savage's enjoyment of existence. Take the lowest savages known to us, the native tribes of Central Australia, and turn to the most recent and the best accounts of their manner of life; and it is certain that their existence is enjoyed by them. Is ours without exception enjoyed by us?

If it is easy to be led by sentimentalism into mistakes about what our fellow-man thinks of the question whether life is worth living, it is still easier to be misled with regard to our fellow-creatures lower in the scale. Here all is conjecture, and it is on this uncertain ground that rests the charge brought against Nature of waste and cruelty. There is the cruelty with which, in order to secure the survival of the few and fittest, the environment kills off the many who are unfit--an argument of great force, if the survivors were immortal. There is the waste of bringing into life thousands of creatures unfit, and therefore doomed to a speedy extinction. But death is the common lot; and as for waste and failure, if the short-lived creatures fulfil their purpose, they are not failures; and if their purpose is by competition to force the development of the potentially fit, then they fulfil their purpose. A man may be entered for a race for no other purpose than to force the pace. As for happiness, wild animals, to judge by their usual fit condition and by the evidence of sportsmen, do enjoy existence. But they, at any rate, have no ideals--whatever the savage may have. Yet it is conceivable that the bird that builds its nest finds some satisfaction in doing so, and that the animal that lays down her life to save her young ones has some sense of love. What is revealed as the ideal in man may be inchoately manifested as instinct in the undeveloped consciousness of the animal. If so, then the animal's life has independent value and is not merely valuable as a means to a distant future end.

To sum up: science declines to take the teleological view of Nature, or to admit final causes or ends. To speak, therefore, of survival in the struggle for existence as an end, may be excellent sense, but it is unscientific: it implies an assumption of a kind about which science is agnostic. If we do, however, make this one deviation from agnosticism, we have then no difficulty in showing that evolution is a failure, for its end is survival, and we all die; and there is no compensation, or, if there is, posterity gets it, not we--an aggravation of the original injustice.

If survival in the struggle for existence is the only end that we personally recognise in the conduct of our own lives, we are quite consistent in judging it to be the only end of other lives, and in condemning the Universe, for then there is neither goodness nor any wisdom in it.

On the other hand, our faith in that wisdom and goodness is not genuine so long as we are prepared to stake only our arguments on it, and not our lives.

XIII.

EVOLUTION AS PURPOSE

Evolution, as a scientific theory, is a description of the process by which the totality of things has come to be what it is. The method employed is that of science, and proceeds upon the assumption of the uniformity of Nature and the universality of the law of causation. The existence of a thing is proof that the conditions necessary to produce it preceded it. Thus from what is we infer with certainty what has been: the occurrence of Z is proof that Y preceded, and so from Y we can infer X, and so on, to the beginning of the alphabet. Eventually, that is, we are carried back, in theory at least, to an initial arrangement of things which not only gave birth to the actual order of evolution, but was such that no other order of events could have followed from it. Were it possible, in fact, to get back to this original collocation of causes and to formulate it, the formula would explain the universe as it is and has been, the totality of things.

Unfortunately the formula, though it would explain everything else, would not explain itself, and would therefore, so far, fail to explain anything. Or, to put it in other words, though certain causes, collocated in the proper way, would, on this view of evolution, explain everything which ensued from that collocation, we should still want to know why the causes were collocated in that particular way rather than in any other. To say that that collocation was not the outcome of a previous collocation is really to say that there was originally no antecedent necessity why this or any other order of evolution should take place at all; that Z hangs on Y, Y on X ... and A on nothing at all; that the formula which is to render all things intelligible is itself unmeaning. Or, if we say that things had no beginning--matter and force being indestructible--then there is no initial collocation, that is to say, no formula, even in theory, to explain all things: we cannot even imagine the process of evolution to be intelligible.

The latter seems to be preferred by science as the final result of scientific knowledge: the object of science is to demonstrate, not why, but that things happen in a certain way; and it is admitted, or rather insisted upon, e.g. by J. S. Mill, that if scientific knowledge were carried to its utmost conceivable or inconceivable perfection, the question why anything should happen or does happen would remain as great a mystery as ever, and must remain so, for the simple reason that it is a question which science does not even put, much less attempt to answer. Nevertheless, it is said, science does prove what she undertakes to show, viz. that things do happen in certain ways, which ways when formulated appear as laws of science. That, however, is not strictly the case if science, in order to prove her conclusions, has to postulate that each and every state of things is the outcome of some antecedent necessity. Ultimately the postulate proves untrue; for there can have been no necessity antecedent to the initial arrangement of things. And if the postulate be untrue, the conclusions based on it cannot be accepted as certain. If we cannot tell whether it be true or not, neither can we tell whether science be true or not. If it is unintelligible, no wonder that things, as explained by it, are mysterious.

But let us waive these theoretical objections. Does Science, as a matter of fact, prove that things do happen in the ways she describes? In justice to her, let us remember that she does not undertake to do even that. Her laws only state the way in which things tend to happen, not the way in which they actually do happen; only what would happen if there were no counteracting causes and if certain conditions, which do not prevail, did prevail--not what does really happen in the world as we know it. Herein the scientific reason behaves in exactly the same way as the moral or religious reason. Science no more alleges that all bodies in motion do move for ever in the same straight line, at the same rate, than the moral reason alleges that all men always do what is right or that they always do God's will. The allegation is that the tendency exists and can be discerned by those qualified to form an opinion on the matter. That there is friction retarding the movement, and that there are obstacles diverting it, is admitted; and, though the admission does not affect the truth (in one way) of the laws of science, it does allow that they convey no exact or faithful picture of what actually happens in the world as it is.

But if the laws of science do not explain what happens--even in the limited sense of scientific explanation--they are the indispensable preliminary to that explanation. If they do not represent the world as it is, they supply the means by which we may hereafter produce the picture. They are ideals not in the sense that science hopes to show eventually that feathers only appear to float leisurely to the ground, and are really all the time falling sixteen feet in a second, but in the sense that starting from the gravitation formula we could show that every feather's fall is as rationally comprehensible as the gravitation formula itself. They are not the ultimate truth, the final reality, or Science's supreme ideal. They are shadows cast by the scientific ideal before its coming; they are the principles by which science must proceed, if she is to make the world of things intelligible. So, too, the reality of the moral ideal does not imply that, in refusing to make sacrifices for others, I only appear to be selfish, and shall be found in the end really to have been actuated all the time by some high moral principle. What is implied is that only by the acceptance of the moral ideal can the world of men be moralised. From the same point of view it seems hopeless to try to make out that atheism, though not in appearance, will be found in reality to have been a manifestation of religion. It is by accepting, not by denying the religious ideal or doubting its existence, that the ideal of religion is achieved.

Now, in the theory of evolution we have the attempt made to effect this transition from the abstractions of science to the concrete facts, to show that the world as presented to sense is as intelligible and rationally comprehensible as the laws of science themselves, and that the hypothetical statements of science were but preliminary, though necessary preliminaries, to a categorical statement of actual facts. In evolution, as indeed in all the historical sciences, we abandon the elasticity and the uncertainty of conditional conceptions for the rigidity and certainty of accomplished fact. We no longer deal with what may happen if given conditions are realised, but with what has been, and therefore is subject to no "ifs." We start from the certainty of what is, and thus we argue back positively to what must have been.

There is, however, one precaution which must be observed, and without which the whole of the system just described is as uncertain and conditional as the rest of science. Before we can argue from what is to what has been, we must first know for certain what is. Before we can conclude that a patient has been healed by faith or cured miraculously of an incurable complaint, we must first have medical evidence to show that he had the disease. Or, to take a better and closer illustration from medicine, it is premature to assign a cause for a patient's condition before his condition has been diagnosed; and physicians who differ in their diagnosis will naturally differ as to the causes in the patient's past history which are responsible for his state.

If, then, the evolutionist is to attain accuracy in his description of the process by which the totality of things has come to be what it is, he must first know what it is. Before we can trace the evolution of morality, for instance, we must make up our minds as to what it is. If we regard it as an illusion, we shall hold that it is subject to the same laws as other illusions, and we shall have no difficulty in showing that its evolution was a necessary consequence of those laws. Or, again, if we hold that religion is mere foolery or hysteria, we shall naturally infer a very different process for its evolution than if we feel it to be a permanent manifestation of the common consciousness in the same sense that morality is. A distinguished German mythologist, starting from the former diagnosis, has no difficulty in evolving primitive religion out of primitive drunkenness.

In fine, if we regard "what is" as giving the data by which we are to determine what has been, it is clear that to understand what has been we must properly appreciate what is. This is in accordance with the conclusion which we have reached previously that it is only by studying its effects that we can properly understand a cause. To judge a thing properly we must know the effects it is capable of producing: to know what a thing is we must observe what it becomes or what it is capable of becoming at its best. We cannot judge the value of the moral character or the moral ideal fairly if we take a low specimen to go by; nor if we knew nothing more of morality than what we could observe of its rudiments in the higher animals, should we know much about it. It is by its highest manifestations that we most correctly judge either morality or art, and it is only through them that we can be properly said even to understand what either art or morality is. So, too, taking the religious ideal as love of God and man, we must judge religion not by its imperfect manifestations in imperfect beings, but by its perfect revelation and realisation in Christ.

The case is not otherwise with science or evolution itself. From primitive times man has always used his knowledge (however imperfect) of what is as the basis of speculations as to what has been. It would, however, be absurd to take the puerile and barbarous cosmogonies of the savage as adequate expressions of the scientific ideal, or to imagine that it is from them that we can judge what science is. It is no less unreasonable to judge the theory of evolution by its present, passing phase. In the first place, there are facts in its history which show that it naturally started with a partial and one-sided view of the facts. In the next place, we must judge it not by what it may be at its worst, but by what it is capable of becoming at its best; and it is by the latter that we must decide what evolution truly is, not by the former.

At its worst the theory of evolution may require us to believe that the whole process of evolution is essentially irrational--being the outcome of unintelligent forces operating on reasonless matter--and that the theory of evolution, accordingly, if faithful to the facts, is as irrational as they; or, if rational, is a misleading account of the real universe in which we live and move and have our being.

On the other hand, the theory at its best may require us to believe that it reveals a universe run on rational principles, a real world perfectly intelligible to perfect reason, and partially intelligible even to beings who share but partially in the Divine reason that animates the whole.

Both theories, however, base themselves upon what is, and profess that their conclusions follow logically from it. If, then, they differ in their conclusions it is because they differ in their diagnosis of what is. Both admit the existence of faith; but one regards faith as a fact in the pathology of human reason, the other regards it as the normal mode of our common reason's operation. The latter, therefore, requires to postulate causes which will account for the correctness of the common faith of mankind; the latter, causes which have resulted in the common illusion of mankind.

It seems, then, that even in evolution we do not escape after all from the indeterminate and conditional knowledge, which science offers, to the absolute certainty of accomplished fact. Every theory of the past history of the world is just as conditional, just as much dependent on an "if," as the hypothetical laws of science, for any such theory is dependent on the view it takes of what is, and is correct only if that view is correct.

The theories of evolution which we have called the Optimistic and the Pessimistic interpretations of evolution are avowedly based on the assumption that a large part of the common faith of mankind is a mental or moral disease. According to Mr. Herbert Spencer the faith that we can know what is real is an illusion: the Real is the Unknowable. According to Professor Huxley the common faith in the freedom of the will is an illusion: necessity is the law of the uniformity both of Nature and of human nature. In thus declining to accept the testimony of the moral and religious consciousness as evidence of what is, both philosophers were influenced by the belief that it is science alone which is capable of ascertaining and demonstrating what is and what actually does happen. This belief, however, we have ventured to suggest, overlooks two facts. One is that the abstract sciences do not even profess to state what actually does happen: they simply affirm that, if the conditions stated in their various laws are the only conditions operative, the only result will be that stated by the particular law in question. Thus science does not concern itself with what is or does happen, but solely with what would be or would happen under certain (usually impossible) conditions. The other point overlooked is that the historical or comparative sciences are also only hypothetically true. All that their laws undertake to demonstrate is that, if certain consequences constitute the whole of an observed effect, then the only conditions antecedently operative were those stated in the law. Here too, then, science does not even claim to prove what is or demonstrate what does happen, but assumes that we know it or find it out, in some way with which science does not concern itself. If we do know and can know what is, science can tell us what were the conditions that produced it.

The question, then, that we have to put to any theory of evolution--that is, to any theory which professes to state the process by which the totality of things has come to be what it is--is, "Does it account for that totality? do the causes which it assumes to have been at work account for all that is?" Now, a priori it was not to be expected that evolution would in its infancy, and it is still young, succeed in accounting for all things; and there were special reasons in the circumstances under which it first took its modern scientific shape which necessarily limited its earliest attempts to grasp the totality of things. It would, however, be absurd to judge the principle by the first attempt to apply it, and to condemn it because it has not done in a moment what with time it assuredly will succeed in effecting. At the same time, it can only effect that wider success by refusing to stereotype its first errors and by declining to bind itself to the dogma that what it has succeeded in explaining is all that there is to explain, or that that alone is or happens which its present assumptions or laws are capable of accounting for. There lies the danger which threatens to check the further development of the theory of evolution--in the dogmatism which pretends to set aside common sense and the common reason, and arrogates to itself the sole right of saying what is; and succeeds in doing so by the simple but circular argument that that alone is or happens which can be accounted for by the laws that regulate the movements of things in space or that follow from the struggle for animal existence.

Historically, the theory of evolution in its first manifestation was an extension to the historical sciences generally of a purely biological conception, that of the origin of species as a consequence of the struggle for existence. It was found that much else in the manifold of what is, many other differences between related things, besides the differences which mark off one species of animals from another, might be accounted for, historically, by the theory that those differences were but the sum and the accumulation of an infinite number of small modifications which had given the thing an advantage over its rivals in the struggle for existence. Strictly speaking, all that this remarkable and wide-reaching discovery implied as a matter of logic was that between animals and things not animal there existed an analogy or resemblance, in virtue of which it was logical to argue from things animal to things not animal just so far as the resemblance between them went, but not further. Very naturally, however, it happened that with this originally biological conception all its biological implications were taken over, and it was (and is) argued not merely that there are great and fruitful resemblances between, say, society and an animal organism, but that societies are animal organisms. In fine, sociology was treated as a department of biology. The fallacy that science demonstrates what is, and that what science does not account for has no real existence, thus made its appearance simultaneously with the birth of the evolution theory. The resemblances between the evolution of the social organism and of animal organisms could be accounted for by the biological theory of the struggle for existence; the differences, therefore, must be denied or laboriously explained away. With the growth of sociology, however, it is becoming apparent that the evolution of society has laws, some of which do indeed coincide with those of animal evolution, but others of which are peculiar to sociology in the same sense as the laws of chemistry are distinct from those of physics. Sociology is accordingly revolting from its bondage to biology: the plain fact that society is not an animal is beginning to make itself felt. The resemblances between the organisation of society and that of an animal are freely admitted, but the differences are beginning to claim consideration also; and the sound doctrine is beginning to assert itself that by experience alone, experience of what is, and not by any a priori dogmatism as to what in the name of science must be, can we tell how far the resemblances extend as a matter of fact and where the differences begin. That the evolution theory must be the gainer by thus admitting the facts instead of denying their existence is clear; if sociology is not a branch of biology, and yet the two sciences have certain laws in common, a great step is at once taken towards demonstrating the existence of certain general principles of evolution which are higher than the laws of either, or perhaps than of any, particular science.

The tendency of the scientific theories prevailing for the moment to deny the existence of what they cannot, for the moment, account for, is exemplified in another way by the theory of the survival of the fittest. It was shown by Darwin that, granted the tendency to variation in animals, the struggle for existence was enough in its results--as he had the genius to discern them--to account for the origin of species. The struggle for existence is a fact, and thus animal evolution was based on what is, on positive fact. To apply the same process of argument to human and social evolution was perfectly scientific and legitimate. What is neither scientific nor legitimate is to maintain, explicitly or implicitly, that the totality of human activity is engaged and exhausted in the struggle for existence. Self-preservation is undoubtedly a powerful instinct, but it is not the only instinct even of animals, and is not always the most powerful in man--or in the brute. That there are resemblances between man and his fellow-creatures, the brutes, and that so far as those resemblances extend, man and the animals have been, and are, subject to the same laws of evolutions, are facts which may be heartily admitted, but which neither authorise us to deny the existence of specifically human peculiarities, nor warrant us in trying to deduce the differences from a law which applies only to the resemblances. If the evolution theory is to state the process by which the totality of things has come to be what it is, it must begin by facing the whole of the facts--in this case by admitting that not only have the fittest to survive survived, as is natural in a struggle for existence, but that progress, æthetic, ethical, and religious, has been made.

The denial of this fact may either be open and avowed, as, for instance, when the reality of the religious ideal is formally denounced; or it may be tacit and implied, as, for instance, when moral progress is defined as adaptation to environment, i.e. as not progress at all, or when the freedom of the will is denied, i.e. when approximation to the ethical ideal is maintained to be a thing not under our control. Tacit or avowed, this denial proceeds upon the fallacy that the laws of science, as understood and formulated at any particular moment, are the sole test and constitute our only knowledge of what is. But the interests both of the common sense of mankind and that specially organised form of common sense which we know as science require a protest against that fallacy: it is opposed to the principle on which scientific knowledge rests, and it would be fatal, if acted upon, to all further development of that knowledge.

The principle upon which science rests is that its laws are capable of verification, and that they are verified when and if they are confirmed by experience. The final appeal of science is to the evidence of consciousness, the only evidence of what is that we possess: the only evidence of the truth and accuracy with which an eclipse has been calculated is the evidence of our senses that the eclipse does take place and is visible in the place and at the time predicted. If a hypothesis predicts results which as a matter of observation do not take place, the hypothesis is judged so far inaccurate or inadequate: what is over-rides our preconceived opinions, even if they be the hypotheses of science, as to what ought to be or will be. It is the ever-open appeal to the final court of fact, of what is, that condemns false assumptions, guarantees the truth of science, and safeguards the freedom of scientific inquiry. To allow any group of men, however eminent, or any body of science, however sound, to deprive us of this right of appeal and bid us disbelieve in the evidence of our own senses, if it contradicts their theories, would be to submit to the tyranny of dogmatism, and to be faithless to the cause of truth.

Fortunately, though the unconscious and therefore ill-considered metaphysics of some men of science have tended in the direction of scientific dogmatism, the practice of science has been in the opposite direction. In practice science has owed much of her progress to the study of "residual phenomena." Phenomena which the laws of science for the moment could not account for have not been denounced as illusions, or ruled out of court as non-existent or beneath the notice of science: they have been accepted as facts, as part of the totality of things which it is the ambition of science to account for; and, accepted as such, they have led, it may be, to the discovery of a new planet or a new element, but always to the discovery of fresh truths, which never would have enriched the page of science had science refused to take cognisance of facts the laws of which it had not at the time discovered.

In demanding, then, that any theory which professes to account for the totality of things should recognise the fact of ethical and aesthetic progress, and that all progress is willed and purposed, we are seeking not to cramp science but to enlarge its bounds, not to introduce a new scientific method, but to extend the application of existing methods, and to carry out the principle on which the truth of science and the freedom of scientific inquiry are based. The laws which enable the physicist to explain the mechanical action and reaction of things do not suffice to explain the reactions studied by the chemist. The laws of chemistry are inadequate for the purposes of the biologist. It is but an extension of the same principle when the student of the anthropological sciences finds it necessary to assume, or rather discovers, that the laws of animal existence do not wholly account for everything that man does; and it is to these sciences that we must look for the next important and fruitful modification of the general theory of evolution. It is to them, dealing as they do with the highest product of evolution, that we must look for the truest interpretation of evolution. On the principle that to understand what a thing is we must not reduce it to its lowest terms, but look at it in its highest manifestation, we must judge the evolution process by its highest phase, by all that it is capable of, and not by the least we can, by scientific abstraction, leave in it. And the sciences which, merely to maintain their scientific existence, have a vital interest in insisting on the reality of will and purpose as causes which have influenced the direction of the evolution process are the sciences which deal with man.

Those who find it easy to believe that a society is an animal, like those who proclaim that the real is unknowable, but that our knowledge of it is just as good as if it were not unknowable, will have little difficulty in believing that men's actions are not influenced by their purposes; and both will probably subscribe to the doctrine that, first, approximation to the ideal is an unintended result of the brute struggle for mere animal existence; and, next, the purpose which appears to mark the evolution process and to be the cause of progress is semblance only, a mere illusion. Against the first article of this doctrine the final and decisive appeal is and always must be to experience. It makes a general statement with regard to particular facts of experience: like every other statement made in the form of a scientific law, it affirms that a certain proposition will be found, when tested by experience, to be true of every one of a certain class of facts in our experience. It is therefore competent for every man, who chooses to consult his experience, to decide for himself whether the statement is true. In the present case, it is for every man, who has struggled with temptation and has achieved any progress, to say whether he gained the victory without an effort of will, without any desire for better things, without any purpose or resolution to try once more, without any intention not to yield the next time. Are "secret commissions" in trade refused, when refused, unintentionally? or is their refusal due solely to the blind instinct of self-preservation in the struggle for commercial existence? If reform is effected, will it be effected by those who declare that the severity of the struggle for existence makes reform impossible? or by those in whom the ideal of honesty has some operative force and who purpose approximation to that ideal? When the conviction is expressed that public opinion alone will be able to check this form of dishonesty, what is that but an appeal to the common sense and common faith that there are other things which man can will and purpose besides success in the struggle for existence?

The doctrine that the universe presents the mere semblance of purpose, that Nature mimics purpose, having none, is shared by materialistic systems in common with all those which consider that the only explanation that can be rendered of any given state of things is the assumption that it is the issue of some antecedent necessity which produced it. As we have already argued, the assumption of necessity as the ultimate explanation of things breaks down when we come to consider the beginning of the universe. If we assume an absolute beginning, then there can have been no necessity antecedent to that, and the beginning of things is left without explanation. On the other hand, to say that there never was any beginning is to admit that there never was any original necessity why things should follow the course of evolution which they have pursued--the initial collocation of causes was due to chance, was a purely fortuitous concurrence of atoms. When it is remarked that this is a strange assumption, that really, if the whole evolution process had been designed to reach the stage in which we know it and to attain the ideal which we surmise it to be capable of, the primeval atoms could not have been arranged better for the purpose, the reply is that the appearance of purpose is a delusion: true, as a matter of chance, the chances are millions to one against a fortuitous concurrence of atoms producing the evolution process that has taken place, but then the chances were just as great, neither more nor less, against any other of the millions of evolution processes that might have been evolved. We know the one that has taken place, and it is marvellous in our eyes that precisely this and no other should have occurred; but the wonder vanishes when we reflect that, had any other occurred, we should have been equally convinced, and equally erroneously convinced, that it could not have been produced by chance. The initial arrangement of things was, as it happened, such as to produce our evolution process: things might have chanced differently at the beginning; if they had, a different evolution process would have taken place, that is all. But it would still have looked like purpose, and would still have been due to chance.

But would it? The whole question is whether the initial collocation was due to chance or to purpose. To say that there might have been many other collocations proves nothing: an Almighty Power could collocate things in any of an infinite number of ways. To argue that every possible collocation, and therefore the one that produced our evolution process, must be due to chance, is simply to beg the question: the very thing we want to know is whether this or any other process could be due to chance. The argument that any and every other process would equally testify to purpose and equally imply design, seems rather to indicate that no conceivable evolution process could conceivably be due to chance.

Next, the necessitarian argument lays it down that the marvel of evolution vanishes when we reflect that if things had been different at the beginning, the results would have been different. But they were not. And the fact that they were not is just the marvel which the necessitarian does not even explain away: in order to diminish the probability of purpose, he postulates countless possible alternatives to the original arrangement of atoms, and then he is embarrassed with the difficulty of getting rid of them. Why was this particular collocation determined on rather than one of the countless alternatives? To say it was chance may be true; but we want to know what reason there is for believing it to be true. If there is none, then neither is there any reason for believing the purpose that makes the evolution process to be an illusion.

But let us grant it was chance: chance, as everyone knows, is merely a name for our ignorance as to the real cause; so that to say it was due to chance is to say that, for anything we know to the contrary, the original concurrence of atoms may have been due to purpose. In a word, there is, on the theory of chance, no reason to believe that purpose either is or is not an illusion.

It may, however, be said that not only do we not know, but that we cannot know, whether it is an illusion or not. In reply we may either admit that all our knowledge--scientific, moral, and religious--is based not on knowledge, but on faith; or we may ask on what grounds this alleged impossibility is based. If we put that question, we shall find that the grounds are not altogether cogent. It is alleged to be equally impossible for the human mind to conceive either the existence or the non-existence of a necessity antecedent to the absolute beginning of things: therefore, in face of this inherent incapacity of the human mind, the truth about the beginning of things is unknowable and inconceivable. But, we venture to suggest, this alleged incapacity of the human mind rests on a false antithesis: it rests on the assumption that whatever phase of the evolution process we regard as the initial arrangement must either have been determined by some prior phase (in which case it was not initial) or not determined at all. But as a mere matter of logic, there remains the possibility that it may have been self-determined; and, as regards the evidence of experience, we are familiar with a cause which operates every day and which is self-determined, viz. the free will. There is, therefore, no such inherent incapacity in the human mind as is alleged; and the only inconceivability is that which is inherent in the theory of antecedent necessity, and not in the facts themselves. It is simply incorrect to say that if things cannot be explained by the theory of antecedent necessity, they are not capable of being explained at all. If the evolution process had been designed to follow the course it has followed, the initial arrangement of things could not have been better adapted to produce the result; and, as adaptation of means to end is the mark of intelligence, it is neither inconceivable nor irrational to suppose that purpose was immanent in things from the beginning.

But as it is scientific to argue from the known to the unknown, or from the better known to the less known; and as to know fully what a thing is we must know what it is capable of becoming or producing, let us pass from the pre-animal to the animal stage of evolution. It is the more necessary to do this because it was Darwin's theory of the origin of species which impressed upon the modern mind the idea that Nature mimics purpose, having none. Man, with the purpose of breeding a certain type of animal, selects those animals to breed from which possess, in the most marked degree, the characteristics which he wishes to develop in the offspring. But, as Darwin demonstrated, Nature, or the environment, by killing off those creatures which did not possess (or least possessed) the qualities necessary to ensure survival, "selects" animals of a certain type to breed from. Thus "natural selection" produces its results in the same way as human selection does; and presents every appearance of purpose, though the environment which produced the results could have had no intentions or purpose at all. But just as man does not create the animals which he first selects to breed from, so the environment does not create those sports or varieties which it selects to breed from: if they did not exist, neither man nor Nature could breed from them--no results, purposed or unpurposed, could be got from them.

If now we inquire about these sports, we are told science is content with the fact that they undeniably occur: wherever there are animals there are varieties in their offspring. That those which are adapted to survive will survive, and those which are not will not, is a self-evident, indeed an identical, proposition. It is; and it gives away the whole case against purpose, for it admits that some varieties are originally adapted to survive, that without them neither man nor the environment would have anything to begin on or work on, and that though man and Nature may develop, they do not create the original adaptation. They do but promote, by conscious or unconscious action, the purpose immanent in the sport. Of all the numerous, successive, imperceptible increments by which what was originally a sport is raised to a distinct species, not one is created by man or by the environment: all are the "gratuitous offerings" of the organism, manifestations of the organism's spontaneity, revelations of its latent capacities, fulfilments of the purpose immanent in it from the beginning.

If it be said that the survival of any or every given species was a matter of chance, because other sports would have developed into other species, if the environment had been different, the reply again is, But it was not; and, on the theory of necessity, could not be. The fact that both conditions--the organism's spontaneity and the environment's selective agency--were requisite to the production of the new species, and that both conditions were forthcoming, tells rather in favour of purpose than against it. The fact that this particular combination of conditions was effected, rather than any other, is on exactly the same footing as the initial concurrence of atoms: if the latter cannot be ascribed to any necessity antecedent to it, neither can the former; the reason of the combination is to be sought in the self-determining cause immanent in the conditions. The fact, if it be a fact, that countless other combinations were possible, and this alone was chosen, shows that the will immanent in the evolution process is free will.

In fine, Darwin has shown that the action of the environment is exactly what it would have been had it been designed for the purpose of selecting certain sports for development. All that is further necessary in order to show that this apparent purpose is an illusion, is to prove that the environment was not designed to act as it does. Pending the production of that proof, the argument remains incomplete.

The larger part of the process of evolution is known to us only from the outside: we observe its effects in the animal world and in inorganic nature, but its inner workings we have to reach by inference. One part of the evolution process, however, we know from the inside--that part which is carried on through us. We are some of the innumerable channels through which the motive force of the process is transmitted; and the knowledge which its transmission through us gives us is more intimate and direct than that which we get from observing the external effects it produces elsewhere. The evolution of society, for instance, is a part of the general process of evolution, and is a process which is carried on through us and expresses the resultant of the totality of our sentiments and actions towards one another. What light, then, if any, is thrown by sociology on the general question of purpose?

Mr. Herbert Spencer has familiarised us with the lesson that in politics and social experiments it is the unforeseen and unintended results of legislation which are far the most important, and that the industrial organisation of the country, or we may now say of the world, is not the fulfilment of any design preconceived by any governmental agency, but the unintended result of innumerable actions on the part of men who never dreamed that their action would have any such outcome. The reason of this is to be sought in the fact that society is an organism and that its growth follows the same laws as those which regulate the structural development of an amoeba or a rhizopod. Thus, both society and the animal organism must be fed. To be fed, both must appropriate nutriment from the environment. That nutriment must be taken up and must be distributed to all parts of the organism, social or animal, if all parts are to be fed--and all must be fed, because all are mutually dependent, and to neglect one would disorganise the whole. Channels of communication must be established between all parts, in order that food may be conveyed from the organ which took it from the environment to the organs which require it for support. What marks the process of evolution in both cases is the increasing division of labour and the increasing interdependence of the parts on one another. The animal organism, like the social organism, is made up of a multitude of living units, each one of which is continually adjusting itself to the requirements of all the rest. The increasing complexity in the structure of an animal organism is possible only because the living units of one part take upon themselves new functions, or devote themselves exclusively to one function, in order to benefit the units of a distant part. If they purposed or were purposed to produce that result, they could not behave differently or better. But this appearance of purpose is mere semblance: the minute cells of an animal organism have no intention of producing even a rhizopod or an amoeba. The explanation of this mimicry of purpose lies in the fact of the mutual interdependence of the parts: no change can take place in one organ of society or of the animal without being transmitted through the whole, just as you cannot remove one of the undermost of a cartload of bricks without more or less disturbing all the rest. But what is true of the bricks or of the units of the animal organism is true of the units of the social organism: what we discover in their action and reaction on one another is the operation, not of voluntary purpose, but of invariable laws of cause and effect.

According to this argument, then, the living units of the animal organism resemble in their action those of the social organism sufficiently to warrant us in arguing from the one to the other, and in concluding that there is purpose in the action of neither. But it is obvious that, if the resemblance is great enough to justify us in arguing from the animal to the social organism, it also opens the way for the argument to travel the return journey, from the social organism to the animal, and to reach the conclusion that there is purpose in both. Let us therefore consider what each of these two opposite conclusions requires us to believe.

On the one hand, before accepting the argument that there is no purpose in the action of the social organism because there is none in that of the animal, we must prove that there is none in that of the animal. But that, as we have already urged, is exactly what has not been proved: the utmost that science claims to prove is that the units of the animal organism do behave in a certain way. That way is exactly the way in which they would behave if they were designed to do so; and science leaves it, so far, a perfectly open question whether they were or were not so designed. The argument, therefore, that there is no purpose in the action of the social organism, because none in the animal, breaks down at the threshold. Yet it is on the unproved and unprovable assertion that the appearance of purpose in the animal organism cannot possibly be due to design, and must therefore be a delusion, that we are expected to deny the evidence of our own experience and consciousness and to believe that we, the units of the social organism, have no purpose in the daily acts by which we extend trade or discharge our social functions.

Thus the surmise that Nature mimics purpose, having none, is a conjecture which, so far as it is applied to the pre-human stages of the evolution process, simply plays upon our ignorance; and which, when applied to that part of the evolution process which is carried on through us, we know to be absurd. On the other hand, if there is such similarity between the laws of the one part of the process and the laws of the other part, it must be as allowable to argue from the part and the laws which we do know to the part and the laws that we do not know, as it is to explain the known by what is confessedly unknown. In other words, if the evolution of the social organism is known to be due to purpose, then it is a reasonable inference that animal evolution, which, we are told, follows the same line and laws, is due also to purpose--and if not to any purpose entertained by the cells of the animal organism, then to that of a Will of which their action is the expression.

It is, however, maintained that the continuous social changes which constitute the evolution of society, so far from being the result of the purpose of any individual or of any government, are frequently the very opposite of what was intended by the authors of the changes, and always are notoriously beyond our power to forecast. But the fact that my plans are modified or diverted by my successors or by my coadjutors does not prove that there was no purpose in my plans, or that there was none in the modifications introduced by my successors. And the total result of our united action and purposes may be something different from what any of us individually intended and yet express a common purpose, which is shown by the result to have been more or less present to all of us. A cathedral begun in the Norman style may have taken generations to build and may end in Gothic; and it will express the ideas common to the several builders, in much the same way that a composite photograph reproduces most distinctly the features in which all the persons photographed coincide, and other features more or less distinctly according to the extent to which they are shared in common by the different subjects. Or, to express the effect of the successive actions of succeeding generations, we may borrow an illustration from the game of chess. It is possible for five players to play, taking it in turns to move, so that every player makes one move out of five, and plays alternately for White and Black. The result, with good players, is a brilliant and well-developed game, which is not the game as purposed or intended by any one of the five players, but as continually modified and improved by each every time that he took it up. When, then, we reflect how many players in the game of life there are even in a small society, we can well understand that, though each has his own way of serving the common purpose, none can forecast the result.

Perhaps it will be said that the chess-players have a common purpose, and the players in life's game have none. The reply is that science assumes they have; science assumes that they play to win in the struggle for existence; and only on the assumption that men have common purposes is it possible to frame any scientific account of their actions. The science of Political Economy assumes that it is a common purpose of men to acquire wealth, and that their actions are determined by that purpose. It then goes on to show that if that is their purpose, then the conditions under which it can be and is effected are of a certain kind, e.g. men must buy in the cheapest and sell in the dearest market. It is not necessary to assume, nor does Political Economy assume, that man can only purpose to acquire wealth, or that he must under all circumstances do so. In the same way it is wholly unscientific in sociology to assume that success in the struggle for existence is either a thing that man must aim at, or the only thing that he can aim at: the soldier dies for his country, the martyr for his faith. The institutions of a nation--legal, political, social, and religious--express the predominant purposes for which successive generations of the community have laboured; and the evolution of mankind is the history of the various degrees of success with which men have realised the ideals which they have purposed to attain. The successive reforms by which progress has been effected have all been purposed, and have all been purposed by men who believed, rightly or wrongly, that in so doing they were serving God and their fellow-men, and that the ideals of truth, justice, equality, fraternity, love, compassion, and mercy express God's will and the Divine purpose.

If, then, the outcome of the pre-human period of evolution has been, as a matter of fact, and amongst other things, such as to prepare the earth for man's habitation and to provide him with a mechanism, physiological and psychological, such that he can use it, if he will, to promote what he considers to be progress and advance, it is not unreasonable for him to regard past phases of evolution as so many steps leading to the realisation of the ideals which he cherishes, in his best moments, as his highest purposes. The continuity of evolution and the unity of its process authorise or even compel him to use that part of the process which is carried on through him as a means to interpret the rest. As, in the game of chess played by five players, each player inherits from his predecessor the game as it stands, and carries on, with improvements or modifications, the scheme which he inherits, so in life each player in turn becomes conscious of the ideals which he too may, or may not, as he wills, carry one step nearer to their goal. It is in the continuity with which these ideals are transmitted through one consciousness after another that the continuity of human evolution consists. We are, or may be if we choose, particles in the medium by which a purpose not our own (save inasmuch as we choose to make it so) is carried onwards to its destination. The medium through which progress has travelled in the past is Nature; the medium through which it is now travelling is human nature. By us the ideal, as it is transmitted through our consciousness, is recognised as implying the presence in us of a purpose higher than our own. Whether in the medium of Nature there is any dim consciousness of the progress towards which the changes in Nature conspire, we know not. But the uniformity of Nature and human nature requires us to see in those natural changes the operation of the same power travelling in the same direction as it does through us. In its passage through us it is made known to us as the object of our highest aspirations; the ideal of purity, of holiness, and love; the God for whom the human heart, mistakenly or not, has always sought, and never sought in vain.

XIV.

CONCLUSION

The Pessimistic interpretation of evolution has taught us the lesson that, if we start without belief in the Divine government of the world, study of the process of evolution will not lead us to discern any Divine purpose in the process. Belief in religion cannot begin without faith in God to start with, just as belief in science or in morality is based not on evidence, but on faith. The question remains whether with faith we can believe that the process of evolution is a revelation of Divine love, and whether man's environment has been evolved in such a way as to promote in him that love of his fellow-man and God which is the religious ideal.

If we look at the structure of society, we see it is based on the fact that man has certain needs--of food, shelter, and clothing, etc.--which can be satisfied more effectually by co-operation and division of labour than by isolated, individual action. The man who earns his own living does so by rendering services for which he is paid: he cannot benefit himself without benefiting others to some extent. That is the law under which he lives, a law not of his own making, nor always to his own liking, but a law inherent in the nature of things, and part of the purpose, if purpose there be, in the scheme of things. As a free agent, man may co-operate with his fellows and take his share of the divided labour, or not, as he wills; but those peoples which have carried the principles of co-operation and organisation furthest have fared best. They have availed themselves of the opportunity offered them, and have survived. The failure of the rest to do likewise has not impeded the fulfilment of the Divine purpose that men should help one another. On the contrary, those who decline to help one another voluntarily place themselves at a disadvantage in the struggle for existence, and are slowly, but surely, crowded out by those who fulfil the Divine purpose less unsatisfactorily, and in consequence tend to inherit the earth.

We have already seen that when a man reaches years of discretion he finds that the physiological and psychological mechanism of which he is now in possession, and for the management of which he is henceforth responsible, has a tendency to run in certain grooves: he has, as a child, been taught and has inherited an aptitude to think and act in certain ways. The same remark applies to the social organism. Before or when the individual awakes to the fact that he is a member of a society, he has already been or is the child of parents to whom he renders obedience, and between whom and himself there exist relations of affection. The evolution of man as a purely animal organism has been such that he begins life with a prolonged period of helpless infancy. Unlike the lower animals, which very soon after birth are capable of providing for themselves, he is for years dependent on others. His prolonged infancy is a prolonged period of plasticity, during which he is moulded into a member, first of a family and then and thereby into a member of society. All the higher animals give their offspring some education, an education as good as they received themselves: in the human race alone do parents give their children a better education than they got themselves. It is, however, not the rising generation alone who benefit by the long period of dependence and plasticity which characterises childhood. It is, of course, true that labour expended on the perfecting of tools and machinery is peculiarly productive, inasmuch as the increased efficiency of the instrument more than repays the greater outlay. But as the workman who produces the tool becomes in consequence of his labour a more skilled mechanic, so the education given by the parent to the child is an education not only of the child, but of the parent, and makes both better fitted to be members of society. It not only secures that subordination of the younger men to the elder, which is necessary for the stability of society and the permanence of the tribe, but it also tempers power with responsibility, responsibility not to some external authority, but to the higher principle within the man.

Thus even in the earliest stage of society the anti-social forces of selfishness and the passions do not operate in vacuo and with nothing to impede them. Society at the very beginning is no tabula rasa: the field is already largely occupied, and the direction of social evolution already largely determined, by that affection between parents and children without which neither society as a whole nor the individual as a unit could come into being or continue to exist. It is an unwarrantable libel, even on savage society, to say that in it the ape and tiger predominate in man: the lowest forms of society survive only so far as there exists more humanity than brutality in the dealings of their members with one another. It is a false philosophy of evolution, not a true acquaintance with the facts of anthropology, which rashly assumes that the morally lowest must have been the only primitive elements in the evolution of humanity. The evil and the good in man have existed side by side from the beginning; unselfish affection, as well as selfish desires, has always been part of the equipment of human nature, though the evolution of the former may be a longer and more difficult process, both in the individual and the race, than the evolution of the latter.

In the race moral progress may be expected with much more confidence than it can in the case of the individual. The mere existence of a society, however simple in structure, is of itself proof that the anti-social forces of selfishness and passion are in it less strong than the instincts of neighbourliness and mutual help. Of competing societies those eventually triumph which are least weakened by internal dissension--that is to say, those societies tend to thrive and extend most of which the members are most ready to subordinate their private ends to the public good. Ultimately it is only by the development of this type of individual character that a society can achieve success; and it is this type of character that the competition between nations develops. But essential as it is to the survival of a society, it is by no means so essential to the survival of the individual in his struggle for existence against other individuals. If, then, society were simply a collection of warring atoms, or if the individual's whole activity were expended in struggling with his neighbour and trying to elbow him out, the type of character essential to the survival of society could never be developed, and society itself could neither come into being nor continue to be. The fact is that men not only compete, but co-operate: society is, and from the beginning has been, an organisation requiring from each of its parts some subordination to the interests of the whole.

As the organisation of society grows more complex, the individual becomes less and less capable of existing independently of society, society becomes more and more independent of the services of any individual member, and both these facts tend to foster the social and weaken the anti-social forces in man. Increasing division and subdivision of labour specialises the function of each member of the community more and more, and so deprives him of the general aptitude for doing all kinds of work which is essential to every man who is, as for instance in a new colony, thrown largely on his own resources. Thus the solitary existence which might be just possible for the outcast from a savage tribe becomes a practical impossibility for the average member of any community that has risen above that stage of social evolution. At the same time the point is reached when no one man is indispensable to the community. Society is made up of units so similar to one another that any one can be replaced by some other, and, as a matter of fact, the place of everyone is at death filled by some successor.

The theory of a social contract, as a historical or prehistorical event in the development of any community, has long been rightly discredited: at no time did a number of men, living solitary lives, have a public meeting and formally contract to live together on certain conditions and for certain ends. Man has been a gregarious, if not social, animal from the beginning. Nevertheless, man has certain needs, desires, and ends which can only be satisfied by means of social organisation, and which are quite as potent in holding society together as if, instead of being tacitly at work, they had been proclaimed aloud in a formal social contract. If through any disease the social organism obstructs, or fails to assist in realising, those ends, the dissatisfaction of the individual and the danger to the state are just as great as if a formal contract had been violated: the disappointment of the normal and reasonable expectations of the members of the community is substantially injustice, and is not altogether erroneously stated to be a violation of the common and tacit understanding on which society is in fact if not formally established. Co-operation in labour does imply some sort of engagement, expressed or understood, that the joint product shall be divided more or less fairly between the joint producers. Unfairness in the distribution of social benefits may be of slow growth, but must eventually result in undisguised resentment--appeal is made openly and consciously to justice, which henceforth becomes the ideal of a section at least of the community, and is recognised as a condition without which a healthy social existence is impossible.

It is thus a monstrous perversion of the plain facts to represent the struggle for existence as having been the sole or the main factor in social evolution:

every member of a community is born into an atmosphere of co-operation and maintains his existence by the co-operation of others. If he must labour to live, he cannot labour for himself without at the same time rendering service to others; the very same conditions which make him desire justice for himself constrain him to maintain justice for the community at large. The social environment is, and has always been, such as to lead man in the paths of justice and to train him for the service of his fellow-man. The units which constitute the social environment are men, beings whose physical, mental, and moral structure is the result of a long process of evolution stretching back to beyond the beginnings of life upon this earth, a process which, assuming it to have had purpose, was designed to include in its effects a creature capable of justice and of love.

The full development of the sentiment of justice has been the work of many centuries. At first, when the community is small and nomad, the idea that a stranger has a right to justice is incomprehensible. Even when with the growth of civilisation provision is made for according foreign merchants and others some protection from the law, the idea that the stranger has the same right to justice as the citizen is neither admitted by law nor entertained as a speculation. Indeed, the law, modest though it be, may be in advance of public opinion and of the practice of officials--witness the extortions practised by Roman governors on the Roman provinces. Eventually, however, public opinion outstrips the law and pronounces that even the colour of a man's skin cannot bar his claims to justice, and that the inhabitants of a country, though they be aborigines, have some rights in it. Finally comes philosophy and pronounces justice, absolute and stern, the one thing needful, the one and only duty which it is within the sphere and function of government to maintain.

Unfortunately for the philosophy which maintains this view, it happens that, just when the authority of justice is admitted by the conscience of civilisation to be paramount, justice as an ideal is recognised to be neither capable of realisation nor absolutely desirable. It is obvious that in the best-regulated even of free communities the amount of justice which can be secured by the action of the law and the intervention of the State falls very far short of the ideal; and the multiplication of laws and State inquisitors, which would be necessary if every form of injustice and wrong-doing were to be punished by the State, would be a remedy, if indeed it were a remedy, worse than the disease. It is impossible to pretend to believe that wealth is distributed according to merit in any existing community, or that any governmental system, even if designed solely with that end in view, could ever determine what a man's merits were, or what his reward should be. Nor is the ill distribution of wealth the only factor of injustice, though it is the only factor with which the State could make pretence to deal: sickness and sorrow, grief and pain--nay, the very capacity for suffering and for joy--are dealt to different men in very different measure. It is plain matter of fact that earthly goods and pleasures are not distributed according to merit; and it is just when man's conquest of Nature has become most complete, when society is no longer struggling for a bare subsistence, when the demand for justice is most fully and unreservedly admitted, that the impossibility of meeting the demand and the danger of failing to meet it become most manifest. The poverty which accompanies progress may in one generation be less than it was in the previous generation, but the extremes of poverty and wealth grow daily wider apart, and the number of those who are poor increases in a growing population much more rapidly than the number of the rich. The danger which this rent in the social fabric threatens to the whole structure of society may be exaggerated, but cannot be denied. The mere justice of individualism which has hitherto sufficed to hold society together, suffices now no longer. The justice which limits itself to the fulfilment of those actions to the non-performance of which a legal penalty is attached, is not the one and only thing needful, nor does its force remedy the numerous cases of undeserved misfortune and suffering which the working of our social and industrial system entails. What heals the suffering and saves it from becoming a festering sore that might prove fatal to society, is that love of man for his fellow-man, which is manifested to the poor by the rich to some extent, but chiefly by the poor. The State can only prescribe and enforce external acts of justice; and the external acts which it prescribes are not the bond which holds or can hold society together. The State, in its attempts to modify society through the individual, is as clumsy as the breeder or the gardener in dealing with animals and plants, and must fain be content if it can modify some of the more prominent external characteristics. Nature is much more searching, and, if slower, much more thorough: the real nature of her work, the true character of the force on which she has made the cohesion of society to depend, becomes obvious at the time when the insufficiency of mere justice for the purpose becomes apparent. Imperfect though man's obedience has been to the commandment, "Thou shalt love thy neighbour as thyself," it is to his obedience that society owes its maintenance.

As a matter of fact, then, strict justice is not and cannot be realised in this world. Even the forces of the social environment which are, to a large extent, under man's own control are not and cannot be so directed by him as to secure rewards and punishments in exact proportion to merit and demerit; while the action of those natural forces which distribute fortune and misfortune, pain and the susceptibility to pain, pleasure and the capacity of enjoyment, is still less under his control and, as far as we can see, is still less proportionate to desert. The fields of the unjust benefit as much as those of the just by the rain from heaven; the labourers who enter the vineyard of civilisation at a late hour receive as great a reward as their predecessors who bore the heat and burden of the day, or even greater; when a tower in our social fabric falls, it is not the guilty who are alone or even specially involved in its ruin. From the time of Theognis, at least, men have inquired with despair how the gods could expect worship when they suffered these things to be; and as long as we look upon life as though we were detached spectators, with no care for it save a disinterested desire to see justice done, it is easy for us to declaim upon the absolute indifference of the cosmic process to man and his deserts. But this detached attitude is purely artificial, and we could not make even the semblance of long maintaining it, did we not unconsciously glide into the more natural, but less warrantable, position of tacitly assuming that our own personal lot would be improved if strict justice were done. But is not our resentment against the injustice of the world partly premature and somewhat shallow and short-sighted? Are we sure we want strict justice? Are we so anxious to have our merits weighed? are they so imposing? Can we pray that we may be rewarded after our iniquities? If society could by some supernatural power deal strict justice to all its members, who would, who could live in it? As a matter of fact--to say it once more--it is not by law alone that society lives, but by love, by the long-patient love of father or mother, of wife or husband, of friend or neighbour, which every one of us has accepted and none has fully requited. Our very hospitals are open to all who need them, to those whose suffering is due to their own negligence, or even crime, and not merely to those whose pain is undeserved. A palpable injustice, worthy of the cosmic process itself! And what excuse, if justice, absolute and relentless, be our highest and worthiest aspiration, can there be for appropriating the reward of honest toil to the often fruitless task of offering to those, who have by their own vice sunk into the depths, one last chance of life and of redemption? The mercy which falleth, like the gentle rain from heaven, alike upon the unjust and the just, must be judged by the same standard that we apply to the cosmic process. We may, like the elder brother of the prodigal son, refuse to see anything in man or Nature but a world given up to gross injustice--persons so superior as to stand in no need of forgiveness and no fear of judgment are able doubtless to judge the world and their fellow-man. But the prodigal himself may, perchance, better understand some of the workings of his father's heart, and trust he sees in the apparent injustice of Nature more instances of that mercy which would not have showed itself to him had justice measured love.

It seems, then, that the "ethical process" and the "cosmic process" are not so absolutely opposed to one another as Professor Huxley endeavoured to make out. Both at times act with a calm disregard of justice. In the one case we know that it is a higher principle which takes the place of justice; and it is a reasonable conjecture that the ethical process, which is one outcome or manifestation of the cosmic process, does but reproduce, in this case as in others, the action of the cosmic force which operates through the heart of man as well as through the rest of the universe. It is at any rate inconsistent to condemn the cosmos for exhibiting that quality of mercy which we rank highest amongst the attributes of man: if we take credit to our fellow-men for that quality, in fairness let us give the cosmos the same credit when it displays the same quality. If, as we assume in this chapter, there is purpose in evolution, let us admit that there is some presumption that it is a purpose of love and of mercy.

As it is by faith in science that men of science succeed in solving problems which, for a time, seem beyond the powers of science to deal with, so it is on faith in religion that the religious explanation of the universe depends for its slow but sure extension. With that faith we may succeed in seeing, to some slight extent, that the unequal distribution of pain, as well as of earthly prosperity, is not incompatible with a Divine purpose in evolution. For that faith we must believe that the suffering and sorrow from which none of us is exempt are not evil, unless we choose to make them so, but opportunities for good. Indeed, without that faith we seem forced upon the same conclusion: the man who devotes himself, his soul, his life to the relief of the needy and the suffering cannot make earthly prosperity his chief good, though, as Professor Huxley has said, he may attain something much better. But if we hold that there is something better than earthly prosperity, can we consistently declaim against sickness and sorrow as the worst of evils, or indict a universe because they are not unknown in it? The Stoicism which lent Professor Huxley the strength to teach that man must to the end declare defiance and resistance to the cosmos--resistance unavailing and defiance doomed to certain failure in the end--might also have taught him that the evil which he calls on us to war against is not in the cosmos; that the enemy of the ethical process has his headquarters not in Nature, but in the heart of man. Pain and sorrow are evil to the sufferer who allows them to make him selfish, and to the spectator who chooses to be callous to his suffering. If our volitions do count for something in the course of things, if we are so far free that we can, in response to Professor Huxley's call, doggedly and repeatedly resist the cosmic process, then it is of our own free will, also, that we do evil when the opportunity of good is offered us. Yet we charge the evil upon the cosmos.

APPENDIX

ON BISHOP BERKELEY'S IDEALISM

When one asserts that a writer is wrong in one of the arguments which he uses, it is well to begin by making sure that he really does use the argument in question. For this purpose it is useful to quote the passages in which the writer uses the argument, and such passages, for my own satisfaction, I will speedily cite from Bishop Berkeley. But first, in order that the reader may know that the interpretation which I put on these extracts is not one peculiar to myself, but is in harmony with the general tenor of Berkeley's metaphysical writings, I will quote from Professor Fraser, who, in his preface to the Dialogues between Hylas and Philonous, states Berkeley's argument to be as follows: "As the common reason of men, tested by their actions, demands the permanence of sensible things, even though they are not permanently present to the senses of any one embodied mind, it follows that the very existence of the things of sense (apart from any 'marks of design' in their collocations) implies the permanent existence of Supreme Mind, by whom all real objects are perpetually conceived, and in whom their orderly appearances, disappearances, and reappearances in finite minds may be said to exist potentially."

And now for Berkeley's own words, (1) In the Second Dialogue between Hylas and Philonous (p. 304 of Professor Fraser's edition), he says, "To me it is evident that sensible things cannot exist otherwise than in a mind or spirit. Whence I conclude, not that they have no real existence, but that, seeing they depend not on my thought, and have an existence distinct from being perceived by me, there must be some other mind wherein they exist. As sure, therefore, as the sensible world really exists, so sure is there an infinite omnipresent Spirit, who contains and supports it."

(2) In the Third Dialogue (p. 325 of Professor Fraser's edition) we have: "Hyl. Supposing you were annihilated, cannot you conceive it possible that things perceivable by sense may still exist?--Phil. I can; but then it must be in another mind. When I deny sensible things an existence out of the mind, I do not mean my mind in particular, but all minds. Now, it is plain they have an existence exterior to my mind; since I find them by experience to be independent of it. There is therefore some other mind wherein they exist, during the intervals between the times of my perceiving them: as likewise they did before my birth, and would do after my supposed annihilation. And, as the same is true with regard to all other finite created spirits, it necessarily follows there is an omnipresent Eternal Mind, which knows and comprehends all things, and exhibits them to our view."

(3) The independent, real existence of things is affirmed with emphasis in the Second Dialogue (ibid., p. 307): "It is evident that the things I perceive are my own ideas, and that no idea can exist unless it be in a mind. Nor is it less plain that these ideas or things by me perceived, either themselves or their archetypes, exist independently of my mind; since I know myself not to be their author, it being out of my power to determine at pleasure what particular ideas I shall be affected with upon opening my eyes or ears. They must therefore exist in some other mind, whose will it is they should be exhibited to me. The things, I say, immediately perceived are ideas or sensations, call them which you will. But how can any idea or sensation exist in, or be produced by, anything but a mind or spirit?"

(4) Finally, in The Principles of Human Knowledge, ?90, in explaining the two senses of "external": "The things perceived by sense may be termed external, with regard to their origin--in that they are not generated from within by the mind itself, but imprinted by a Spirit distinct from that which perceives them. Sensible objects may likewise be said to be 'without the mind' in another sense, namely when they exist in some other mind; thus when I shut my eyes, the things I saw may still exist, but it must be in another mind."

Berkeley's argument in brief, therefore, is that we believe things to be permanent, and must therefore believe in a permanent, Divine mind in which they may exist. The question which I wish to raise is as to this permanence of things; for, if things are not permanent, they cannot testify to the permanence of the Divine mind. I will begin my questionings with the concluding words of the last-quoted passage: "when I shut my eyes, the things I saw may still exist, but it must be in another mind." The expression "the things I saw" would seem to be ambiguous. Are the things I saw the sensations of sight which I had, or are they something different? If they are my sensations, they certainly do not exist when my eyes are closed--things are not permanent. If the things I see are something different from my sensations of sight, then the common-sense Realist would seem to be right, and Berkeley's Idealism must be given up. Let us examine each alternative.

It looks, at first, as though Berkeley himself would say that the things I saw are identical with my sensations of sight: in the third passage quoted above he says, "the things I perceive are my own ideas ... the things, I say, immediately perceived are ideas or sensations, call them which you will." Let us, therefore, see the consequences of adhering strictly to this interpretation of the ambiguous phrase. It will follow in the first place that, unless I can see with my eyes shut, the things I see are not permanent, but do cease to exist when I close my eyes. Next, my sensations cannot exist in somebody else's mind--the fact that you can see the object when your eyes are open does not enable me to see it with my eyes closed. On the other hand, of course, it does not follow that because my eyes are closed nobody else can see anything--only, this does not make my sensations permanent, or prove that they can exist in someone else's mind. In fine, if "the things I saw" are the sensations of sight that I had, then Berkeley's argument from the permanent existence of "things" to their existence in a permanent mind breaks down doubly; for, first, my sensations plainly are not permanent; and, second, my sensations cannot exist in another mind, permanent or otherwise.

At this point it is necessary to note that "existence" has been used in this connection in a double sense: actual existence has been distinguished from potential. It is on this distinction that Mill bases his definition of matter as "the permanent possibility of sensation"; but the distinction is derived from Berkeley, who has, as usual, given the most lucid explanation. In his MS. Common Place Book (quoted in Fraser, i. 325, n. 9), Berkeley says, "Bodies, taken for powers, do exist when not perceived; but this existence is not actual.

When I say a power exists, no more is meant than that if, in the light, I open my eyes, and look that way, I shall see the body." Thus far Mill will go with Berkeley; and thus far both are open to the reproach of not giving a plain answer to a plain question. The plain question of common sense to the Idealist is: Do things exist when unperceived? Does the furniture of my room exist when nobody is in the room to perceive it? To which the Idealist replies that if I go into the room I shall see the furniture--which is perfectly true, but is no answer to the question. There is, indeed, no particular reason why Mill should not plainly answer "No," if it were not for fear of giving a shock to the man of common sense who cannot readily comprehend how it is that the coal in his grate has come to be consumed if the process of combustion has been suspended in his absence. But with Berkeley the case is different: for him the permanence of things and the common-sense belief in that permanence have a value as furnishing an additional argument in favour of a Supreme Mind. But he too evades rather than meets the plain question of the plain man: "Do things exist when no one is conscious of them?" His reply is, "Yes, for the Divine Mind is conscious of them"--which again is true, but is not an answer to the question.

However, the point of immediate interest for our present purpose is to ascertain whether the conception of "potential" existence can lend to things that permanence which according to Berkeley necessitates the assumption of a permanent mind. Now, by the potential existence of a body or thing no more is meant than that if I open my eyes and look in the right direction, I shall see the thing; and the things I see are my sensations, ideas, call them what you will. But that I can see with my eyes shut is beyond possibility of proof--it certainly is not proved by the fact that I can see with my eyes open; and neither is it proved by the fact that other people see things when my eyes are closed. In fine, if the things I see are my sensations, then things cannot have a permanent existence; and no inference as to the permanence of the Divine Mind can be drawn.

We are driven, therefore, to suppose that the things I see are different, partially or wholly, from my sensations. And this supposition seems to be implied by various passages in Berkeley. For instance, he says (i. p. 307), "the things I perceive are my own ideas ... the things, I say, immediately perceived are ideas or sensations, call them which you will," where he seems to distinguish what is immediately perceived (i.e. sensations) from something else. And a few lines before he seems to be inclined to admit the existence of something else than my sensations, for he says "ideas or things by me perceived, either themselves or their archetypes, exist independently of my mind."

The permanence of things is undoubtedly an inference. We find by experience that effects which are produced by causes acting before our very eyes are at other times produced by their causes in our absence: the fire burns in my absence as well as in my presence. Obviously, therefore, the thing which produces its effects when I have no sensations of it must be different from those sensations; and it must be an existing thing, otherwise its effects will be effects produced by a non-existent cause. To say that the unobserved cause in these cases is a possibility of sensation does not mend matters. A possible sensation is a sensation which, as a matter of fact, does not exist and never did. It is a piece of pure imagination; and consequently on this theory the whole past history of the universe is imaginary. Neither are matters mended by denying that there are such things as "causes," and affirming that we only know "invariable and unconditional antecedents." How can a possible sensation, that is, an event which did not take place, precede one which does take place? How can an imagination of my mind have preceded the existence of my mind?

Perhaps it may be said that if Mill's Psychological Theory of Mind and Matter is not satisfactory, neither is the theory of the direct apprehension of reality wholly consistent with itself. It affirms the direct apprehension of reality, yet on examination the direct apprehension turns out to be an inference. Thus: things must have an existence different from our sensations because they produce their effects, and therefore exist, in our absence.

The reply is simple. Unless we believed the effects, which we do perceive, to be real things, we should not infer the causes, which we do not perceive, to be real either. Common sense believes that things continue to exist when we turn our eyes away: their existence beyond the range of observation is an inference from their existence in our observation. Their inferred permanence is deduced from their observed independence.

###

www.ingramcontent.com/pod-product-compliance
Lightning Source LLC
Chambersburg PA
CBHW051806170526
45167CB00005B/1903